Essentials of Statistics
for
Scientists and Technologists

Essentials of Statistics
for
Scientists and Technologists

C. MACK

Reader in Applied Mathematics
Institute of Technology, Bradford

PLENUM PRESS
NEW YORK
1967

U.S. edition published by
Plenum Press, a division of
Plenum Publishing Corporation
227 West 17th Street, New York, N.Y. 10011

First published 1967

Library of Congress Catalog
Card Number 67–17769

Preface

Statistics is of ever-increasing importance in Science and Technology and this book presents the essentials of the subject in a form suitable either as the basis of a course of lectures or to be read and/or used on its own. It assumes very little in the way of mathematical knowledge—just the ability to substitute numerically in a few simple formulae. However, some mathematical proofs are outlined or given in full to illustrate the derivation of the subject; these can be omitted without loss of understanding. The book does aim at making clear the scope and nature of those essential tests and methods that a scientist or technologist is likely to need; to this end each chapter has been divided into sections with their own subheadings and some effort has been made to make the text unambiguous (if any reader finds a misleading point anywhere I hope he will write to me about it).

Also with this aim in view, the equality of probability to proportion of population is stated early, then the normal distribution and the taking of samples is discussed. This occupies the first five chapters. With the principles of these chapters understood, the student can immediately learn the significance tests of Chapter 6 and, if he needs it, the analysis of variance of Chapter 7. For some scientists this will be most of what they need. However, they will be in a position to read and/or use the remaining chapters without undue difficulty. Chapters 8 to 13 contain material which is of value to almost everyone at some time, but the remaining chapters are more specialized in content. The laws (or rules) of probability and the binomial theorem are dealt with in some detail in Chapter 9, for those interested in the fundamentals or in fraction defective testing. (The author and some of his colleagues have found the above order of material to give very good results with scientists and technologists, who have, as a result, grasped the essentials without encountering the difficulties that the formal laws of probability followed by the binomial distribution often give at the start.)

A word to the student

Learn the technical terms, they are usually introduced in inverted commas. You must *do all the examples* (all who have learnt statistics have found this essential). These are largely straightforward applications of the text; they are, by the way, in each section and not collected at the end of each chapter; this is to enable student or teacher to identify quickly the nature of each example, and also to enable an occasional user to try out his understanding of a test on a numerical example (answers are given at the end of the book, p. 158). In addition Chapter 16 consists largely of a collection of problems of practical origin which will acquaint the reader with actual situations requiring statistics (and which are also useful as examination questions). The Appendix contains enough tables for most practical purposes of a not too specialized nature.

The pertinent latest theoretical developments and improvements have been included, among them Welch's test for the significant difference between two sample means (which is *independent* of the true values of the population variances), the so-called 'distribution-free or non-parametric' tests, and my own method of calculating confidence intervals when finding the fraction defective of articles whose parts are tested separately (in other words the product of several binomial parameters).

Acknowledgments

Thanks are due to Messrs Oliver and Boyd, Edinburgh, for permission to quote data from *Statistical Methods in Research and Production* by O. L. Davies. Other quotation acknowledgments are made in the Appendix.

The original results of section 9.9 are quoted with the editor's permission from *The New Journal of Statistics and Operational Research*, published at the Institute of Technology, Bradford.

Finally, I must acknowledge my deep personal indebtedness to those who introduced me to statistics, especially to L. H. C. Tippett and Professor M. S. Bartlett, and to my present colleagues M. L. Chambers and M. Gent, on whose knowledge and advice I have leaned heavily in writing this book, as well as to my scientist colleagues E. Dyson, J. M. Rooum, and H. V. Wyatt who contributed many of the practical examples of Chapter 16.

Contents

1

Introduction or 'What is statistics?'

I do believe, Statist though I am none nor like to be,
That this will prove a war—*Cymbeline* Act II Scene I

Often the best way to understand a subject is to trace its historical development and, pursuing this policy, we note that the word 'statist' first appeared in the English language while Shakespeare was a young man (*c*. 1584). It meant 'a person knowledgeable in state affairs' as might be inferred from the quotation above.

Almost exactly two hundred years later the word 'statistics' first appeared denoting the science of collecting, presenting, classifying and/or analysing data of importance in state or public affairs (note that the ending 'ics' usually connotes a science, e.g. physics). Within a further fifty years the word 'statistics' began to be applied to the data itself and, later, particular types of data acquired particular adjectives. Thus 'vital statistics' came to be used for the figures of births and deaths (from *vita* the Latin for life). Later still, *any* collection of data was called 'statistics' and, so, today the word means either 'data' or the 'science of collecting and analysing data', and whichever of the two meanings applies is determined from the context. Though, perhaps, the very latest use of the term 'vital statistics' to denote the most important measurements of bathing beauties should be mentioned (together with the Duke of Edinburgh's witty remark that 'Nowadays, statistics appears to mean the use of three figures to define one figure').

In this book we shall concentrate more on the analysis of data than on its collection or presentation and this analysis derives largely from the study of quite a different type of heavenly body, i.e. astronomy. In astronomy very great accuracy is essential and the need arose to make the maximum use of all the observations on a star to deduce its 'best' position. The methods developed by Gauss for this purpose (which were based on the theory of dice-throwing developed in 1654 by Pascal and Fermat, the very first theoretical work of all) were later applied to state and biological data, and were greatly extended and developed by Karl Pearson and others from 1900 onwards to this end. Later, agricultural experiments (e.g. the measurements of wheat yields with different strains, with different fertilizers, etc.) led to further important analytical developments largely due to Fisher. From 1920 onwards an ever-increasing flow

1

of both problems and solutions has arisen in every branch of science
and technology, from industrial administration, from economics, etc.
Indeed, it would be fair to say that some knowledge of statistics is
essential to any education today. However, the reader need not be
alarmed at the apparent need to acquire a mass of knowledge, for
the subject matter has been reduced to a few clear-cut principles which,
once grasped, apply to most problems. The aim of this book is to
make those principles clear and also to give a clear idea of their use.

We conclude this introduction with some problems which illustrate
the need for, and use of, statistics. Now, shoe manufacturers are
faced with the problem of what proportion of size 5, size 6, etc.
shoes they should make and with what range of widths, etc. Since
it is impossible to measure everyone in the country a small fraction
or 'sample' of the population has to be measured. These questions
then arise: how many people should be measured; how should they
be selected, what reliability can be placed on the answer, etc.? The
modern theory of statistics supplies answers to these problems. (The
Shoe and Allied Trade Research Association has carried out a
number of such surveys since its inception in 1948 and it may be a
coincidence but I can now buy a good-fitting pair of shoes 'off the
peg' and have lost the feeling of deformity that remarks like 'Rather
a low instep, sir' had given me—my thanks are due to the manufacturers and to SATRA.) Again no manufacturer can guarantee
that every article he makes is perfect, indeed, it is quite uneconomic
to attempt to do so, but he likes to be sure that 95%, say, of the
articles reach a given standard. Again, it is usually uneconomic to
test every article (it may even be impossible, e.g. testing the reliability
of striking of matches). The problems of how many should be
tested and what proportion of these can be safely allowed to be
below the given standard are answered by statistical theory. Similarly, in scientific work, no experiment *exactly repeats* itself. Thus,
suppose we measure the velocity of light several times by a particular
method. Each time we shall get a slightly different answer and we
have to decide what the 'best' answer is. Again, suppose we measure
the velocity by a *different* method; then we shall get a different 'best'
answer and we have to compare this with the first to determine if
they are consistent.

These and other problems will be dealt with in this book, though
we shall not deal with the theory in mathematical detail, but will
outline sufficient to enable a scientist or technologist to appreciate
the scope of each method or test. With this aim in mind we start
with one of the early problems in the development of statistics, one
which is essential to a clear understanding of the subject, namely
the presentation of data.

2

The presentation of data

2.1 Graphical presentation (frequency diagram, histogram)

One answer to the problem of presenting a mass of data in a clear manner is to present it graphically. One very useful form of graphical presentation is the 'frequency diagram'. In Fig. 2.1 the contents of Table 2.1, which gives the wages 'statistics' of a small factory, are shown as a frequency diagram.

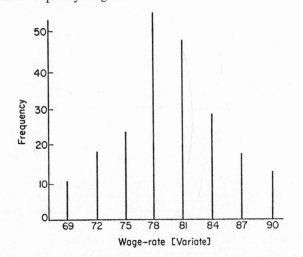

FIGURE 2.1. Frequency diagram showing the wage-rate data of Table 2.1

Table 2.1 WAGE-RATES IN PENCE PER HOUR OF THE MEN IN A FACTORY

Rate in pence per hour . .	69	72	75	78	81	84	87	90
Number of men . . .	10	17	23	55	47	28	18	13

Before describing the nature of a frequency diagram in detail, we define a few basic terms which the reader should know. Data in statistics usually consist of measurements or readings of some property of each person or object in a specified group. Thus, the wage-rates of Table 2.1 are measurements of the earning power of that gróup consisting of the men in the factory. We call the property measured the 'variate', and the number of persons (or objects) who have a given value of the variate we call the 'frequency' of that variate-value. In Table 2.1 the 'variate' is the wage-rate and the number of men with a given wage-rate is the frequency (e.g. the frequency of a variate-value of 81 pence per hour is 47). The essential feature of a frequency diagram is that the heights of the ordinates are proportional to the frequencies.

Example

2.1(i). Draw frequency diagrams for the following two sets of data:

(*a*) *Lengths of words selected at random from a dictionary*

Number of letters in the word . . .	3	4	5	6	7	8	9	10	11	12	13
Frequency . . .	26	81	100	96	89	61	68	42	31	13	11

(*b*) *Observations of proportion of sky covered by cloud at Greenwich*

Proportion of sky covered	0·0	0·1	0·2	0·3	0·4	0·5	0·6	0·7	0·8	0·9	1·0
Frequency . . .	320	129	74	68	45	45	55	65	90	48	676

The frequency diagram gives a good pictorial description of the data if there are not too many different values of the variate (in Table 2.1 there are 8 different variate-values). When there are 15 or more variate-values it is usually better to group some of the variate-values together and to present the results in the form of a 'histogram', as it is called. Fig. 2.2 presents the data of Table 2.2 (this gives the observed heights of 8,585 men) in 'histogram' form. In a histogram the *area* of each box is proportional to the frequency (which, in this case, is the *number* of men with heights within a given box-width).

There are one or two details of presentation by histogram worth remarking on. (1) Where there are only a few observations as with heights less than 60 in. the observations are usually grouped into one box; similarly with heights greater than 74 in. (2) The width of the boxes is usually the same except for grouped frequencies. (3) By the height 62– in. is meant heights of 62 in. or more up to but *not including* 63 in.; similarly 63– in. means 63 in. or more up to 64 and so on. (4) (This point can be omitted on a first reading.) The accuracy of the measurements must be carefully noted—in this case each height was measured to the nearest $\frac{1}{8}$ in.; thus, any height greater than $61\frac{15}{16}$ in. and less than $62\frac{1}{16}$ in. was recorded as 62 in. exactly.

It can be seen from the nature of a histogram that it is less exact than a frequency diagram. Thus we do not know how the 41 heights

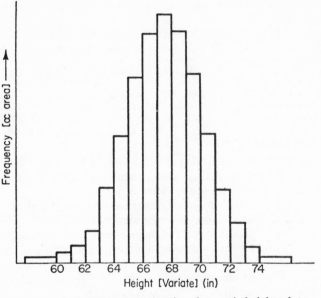

FIGURE 2.2. Histogram showing the men's heights data of Table 2.2

Table 2.2 HEIGHTS OF 8,585 MEN MEASURED IN 1883

Height in in. .	57–	58–	59–	60–	61–	62–	63–	64–	65–	66–	67–
Frequency .	2	4	14	41	83	169	394	669	990	1223	1329

Height in in. .	68–	69–	70–	71–	72–	73–	74–	75–	76–	77–
Frequency .	1230	1063	646	392	202	79	32	16	5	2

in the category 60– in. were distributed; they could all have been $60\frac{1}{8}$ in. though this is rather unlikely.

Frequency diagrams and histograms are still occasionally used in scientific and technical work but the advance of the subject of statistics has been such that rarely are a lot of observations needed and sometimes as few as six are sufficient. However, the idea of a frequency diagram and of a histogram are essential to the understanding of the subject.

Example

2.1(ii). Put the following data into the form of a histogram:

Frequency Distribution of Weights of Adult Males

Weight in lb. . .	90–	100–	110–	120–	130–	140–	150–	160–	170–	180–
Frequency . .	2	34	152	390	867	1623	1559	1326	787	476

Weight in lb. . .	190–	200–	210–	220–	230–	240–	250–	260–	270–	280–
Frequency . .	213	107	85	41	16	11	8	1	0	1

2.2 Some technical terms

When the basic technical terms in a subject have been learnt the student usually finds that he understands everything quite clearly. Here are a few terms he should learn.

The complete set of measurements made on each of a given group of persons or objects is called a 'population' (note carefully that in statistics a 'population' is a set of *measurements* or *observations* and *not* a set of human beings or even animals). The term derives from the state-figures origin of the subject. However, there is an alternative, but less commonly used, word for 'population' namely 'universe', deriving from astronomical origins.

Again, by an 'individual' from a population we mean a single *measurement* or *observation* and *not* the person or object measured. The word 'observation' is itself an alternative word for 'individual', the latter term coming from 'state-figures' and the former from 'astronomical' origins.

Usually an 'individual' or 'observation' is a measurement of some quantitative property such as height, weight, strength of acid, income, etc., but sometimes it is a measure of a qualitative property such as colour of hair, eyes, etc.

2.3 Simple quantitative presentation of data; mean, variance, standard deviation

For some purposes it is better to find some typical figure or figures to present a population rather than present it graphically (hence the use of the term 'statistics' for the measurements of bathing beauties; the bust, waist, and hip measurements give typical measurements enabling a good idea of the figure of the beauty to be formed). Again, a look at the frequency diagrams of the wage-rates in two different factories might not tell us if the average worker were better off in one factory rather than the other. But if we calculated the average wage-rate in each factory some comparison, however odious, could be made. In fact the average is the most important quantity in representing a set of observations. However, the word average is not without ambiguity (e.g. the electrical engineer measures the root-mean-square a.c. voltage as the average voltage) so the term 'arithmetic mean' or just 'mean' is used and is defined precisely as follows:

If in a population the possible values of the variate are X_1, X_2, ... X_N and their corresponding frequencies are $f_1, f_2, \ldots f_N$, then μ, the arithmetic mean, is given by

$$\mu = \frac{f_1 X_1 + f_2 X_2 + \ldots + f_N X_N}{f_1 + f_2 + \ldots + f_N} \qquad (2.1)$$

Thus for the data of Table 2.1 we have

$$f_1, f_2, f_3, f_4, f_5, f_6, f_7, f_8 = 10, 17, 23, 55, 47, 28, 18, 13$$

$$X_1, X_2, X_3, X_4, X_5, X_6, X_7, X_8 = 69, 72, 75, 78, 81, 84, 87, 90$$

$$\mu = \frac{\begin{array}{l} 10 \times 69 + 17 \times 72 + 23 \times 75 + 55 \times 78 + \\ 47 \times 81 + 28 \times 84 + 18 \times 87 + 13 \times 90 \end{array}}{10 + 17 + 23 + 55 + 47 + 28 + 18 + 13} = 79{\cdot}73$$

The arithmetic mean μ of a population does not always give enough information for comparison. Thus the men of the Ruanda district in the Congo have a mean height not much different from Englishmen. But they are composed of three races, the giant Watutsi of mean about 7 ft, negroes of mean about 6 ft and a small people of mean about 5 ft (plus, of course, some admixtures). There are consequently a greater proportion of men over 6 ft 6 in. and under 4 ft 10 in. than is the case with Englishmen. So, besides the mean value, some idea of the *spread* of the observations is needed.

This spread is usually measured as follows: the difference between each observation and the mean μ is calculated, this is called the 'deviation' of the observation or, more fully, the 'deviation from the mean'. The square of this deviation is then calculated for every individual in the population and the mean value of this square is calculated. This mean value is called the 'variance' a term derived from the fact that it measures the variation of the observations. The variance, which we shall denote by V, can be calculated from the formula

$$V = \frac{f_1(X_1 - \mu)^2 + f_2(X_2 - \mu)^2 + \ldots + f_N(X_N - \mu)^2}{f_1 + f_2 + f_3 + \ldots + f_N} \qquad (2.2)$$

Thus, for the data of Table 2.1, we have

$$V = \{10(69 - 79 \cdot 73)^2 + 17(72 - 79 \cdot 73)^2 + \ldots$$
$$+ 18(87 - 79 \cdot 73)^2 + 13(90 - 79 \cdot 73)^2\}/211$$
$$= \{1,121 + 1,016 + 559 + 165 + 76 + 510 + 951$$
$$+ 1,372\}/211$$
$$= 27 \cdot 34 \text{ (pence per hour)}^2$$

There is an objection to the variance which can be seen immediately its units are the *square* of the observations units (for the heights of men data of Table 2.2 the variance is measured in square inches). However, the square-root of V *is* in the same units as the observations and is called the 'standard deviation' and is usually denoted by σ (the Greek letter equivalent to s). That is, $\sigma = \sqrt{V}$ or, alternatively

$$\sigma^2 = V \tag{2.3}$$

Thus for the data of Table 2.1 the standard deviation

$$\sigma = \sqrt{27 \cdot 34} = 5 \cdot 23$$

Examples

2.3(i). Find μ and σ for the following wage-rate data (in pence per hour):

Wage-rate	66	69	72	75	78	81	84	87	90	93
Number of men	4	21	43	60	53	37	20	11	3	1

Is the spread (as measured by the standard deviation) greater or less than for the data of 2.1?

2.3(ii). Find μ and σ for the data of Table 2.2, assuming that the average height of men in the class 60– in. is 60·5 in., in the class 61– in. is 61·5 in., etc.

The standard deviation σ is regarded as a more fundamental measure of spread than the variance V, and, in fact, nowadays we always denote the variance by σ^2 and not by V.

Coding of observations. It is often convenient to subtract a suitable quantity from each observation to cut down the arithmetical labour (e.g. we could subtract 60 in. or 60·5 in. from the men's heights of Table 2.2 and thus work with much smaller numbers). Subtracting such a quantity from each observation is called 'coding'. If we find the mean of the coded observations and then add the quantity back to the coded mean we get the true mean μ. For example, suppose we have a population of just five observations 194, 186, 176, 184, 190. By subtracting 180 we obtain the coded observations 14, 6, −4, 4, 10. The coded mean is 30/5 = 6 and the true mean = 6 + 180 = 186.

Examples

2.3(iii). Use coding to find the true mean of the observations: 375, 381, 369, 384, 377, 372, 383, 368, 377.

Further, the variance σ^2 can be calculated from the *coded* observations and the coded mean as though they were the true observations. This can be seen from the fact that, since

$$X_1 - \mu = (X_1 - A) - (\mu - A), \ X_2 - \mu = (X_2 - A) - (\mu - A),$$

etc., we can rewrite (2.2) thus

$$\sigma^2 = \frac{f_1\{(X_1 - A) - (\mu - A)\}^2 + \ldots + f_N\{(X_N - A) - (\mu - A)\}^2}{f_1 + \ldots + f_N}$$

(2.4)

whatever the value of A. Thus, for the above-mentioned population 194, 186, 176, 184, 190 we have

$$\sigma^2 = \{(14 - 6)^2 + (6 - 6)^2 + (-4 - 6)^2 + (4 - 6)^2 +$$
$$(10 - 6)^2\}/5 = (64 + 0 + 100 + 4 + 16)/5 = 36{\cdot}8$$

Examples

2.3(iv). Find μ and σ^2 for the data of 2.1 by subtracting 80 pence per hour from each variate-value.

2.3(v). Find the variance σ^2 for the data of 2.3(iii).

Sometimes it is convenient to *multiply* observations which are in decimals by some number B, say, which brings them to whole numbers; or, if the observations are very large they may be divided by C, say. The combined operation of subtracting A and multiplying by B (or dividing by C) is the most general form of 'coding'. In such cases the standard deviation calculated from the coded observations will have to be divided by B (or multiplied by C) to get the true standard deviation σ.

2.3(vi). By subtracting 0·0050, or other number, and then multiplying by 10,000 find μ and σ for the following population: 0·0063, 0·0054, 0·0055, 0·0059, 0·0071, 0·0042 cm.

2.3(vii). Find the variance of the population: 17,600, 19,800, 21,200, 23,700, 19,900, 18,700, 20,500, 19,800, 20,300.

2.4 More detailed quantitative presentation of data

The mean and the standard deviation give a sufficiently good idea of the population for many purposes but in other cases more detailed description can be given by finding the mean value of the cube of the deviations from the true mean μ, the mean of their fourth power and so on. Not much use is made of means higher than the fourth, but they play an important part in the theory. For short the mean value of the rth power of the deviations is called the 'rth moment about the mean' and is denoted by μ_r. It can be calculated from the formula

$$\mu_r = \frac{f_1(X_1 - \mu)^r + f_2(X_2 - \mu)^r + \ldots + f_N(X_N - \mu)^r}{f_1 + \ldots + f_N} \quad (2.5)$$

It will be observed that μ_2 the second moment about the mean is the same as the variance σ^2. The mathematically inclined reader will be able to prove without difficulty that

$$\mu_1 = 0$$

Sometimes it is more convenient to use the mean of the rth power of the observations themselves (instead of their deviations). This is called the 'rth moment about the origin' and is denoted by v_r (or μ_r'). That is

$$v_r = \frac{f_1 X_1^r + f_2 X_2^r + \ldots + f_N X_N^r}{f_1 + f_2 + \ldots + f_N} \quad (2.6)$$

There are some relations between the v_r and the μ_r, e.g.

$$v_0 = \mu_0 = 1; \quad v_1 = \mu; \quad \mu_2 = v_2 - v_1^2 \equiv v_2 - \mu^2; \quad (2.7)$$

$$\mu_3 = v_3 - 3v_2 v_1 + 2v_1^3; \quad \mu_4 = v_4 - 4v_3 v_1 + 6v_2 v_1^2 - 3v_1^4 \quad (2.8)$$

The formulae of (2.7) are worth remembering; those of 2.8 are rarely needed.

Examples

2.4(i). Find v_2, μ_2, v_3, μ_3 for the data of Table 2.1.

2.4(ii). Prove the formulae of (2.7) and (2.8) (optional).

2.5 Definition of a sample; its mean and variance

In practice it is usually too costly or time-consuming to measure every person or object in a group, so observations are made on only a relatively small number of persons or objects called a 'sample' of the population. (Note that in statistics a sample is a *number of observations* not just one; in practice a sample may range in size from as few as six to several thousands while in theory a sample may consist of one, two, three up to an infinity of observations.) With samples of small size it is unlikely that two of its observations have the same variate-value and, so, a frequency diagram is not of much value in representing the data of a small sample. The mean and variance of the sample are, however, of the utmost importance and we shall now rewrite for various technical reasons the formulae for these quantities.

Suppose that our sample consists of n observations with variate-values $x_1, x_2, \ldots x_n$ (note that it is possible, though unlikely, that two of these variate values are the same, that is, x_1 could equal x_2; whereas the *possible variate-values* $X_1, X_2, \ldots X_N$ are *necessarily different*). The 'arithmetic mean' or just the 'mean' of the sample is denoted by $\bar{x}$ and its formula is

$$\bar{x} = (x_1 + x_2 + \ldots + x_n)/n \qquad (2.9)$$

The variance of the observations in a sample is denoted by s^2 and its formula is

$$s^2 = \{(x_1 - \bar{x})^2 + (x_2 - \bar{x})^2 + \ldots + (x_n - \bar{x})^2\}/(n-1) \qquad (2.10)$$

It is most important to note that the denominator is $n - 1$ and *not* n as might be supposed from a comparison with the formula (2.4) for σ^2. The square root of s^2, that is, s itself, is called the 'observed' or 'sample' 'standard deviation' (hence the symbol s); we shall call s^2 the 'observed' or 'sample' variance whereas σ^2 denotes the 'true' or 'population' variance.

The use of $n - 1$ instead of n does not affect s^2 much when n is large (if $n = 100$, it makes 1% difference) but at values of n around 10 the difference is appreciable. However, it is the use of $n - 1$ that enables us, often, to get results from as few as 6 observations.

There is an alternative way of writing (2.10) namely

$$s^2 = (x_1^2 + x_2^2 + \ldots + x_n^2 - n\bar{x}^2)/(n-1) \qquad (2.11)$$

which is very useful if desk calculating machines are available, as quantities like x_1, etc. are usually simple numbers but often $x_1 - \bar{x}$, $x_2 - \bar{x}$, etc. are cumbersome decimals. The proof is as follows:

$$(x_1 - \bar{x})^2 + \ldots + (x_n - \bar{x})^2 = (x_1^2 - 2\bar{x}x_1 + \bar{x}^2)$$
$$+ \ldots + (x_n^2 - 2x_n\bar{x} + \bar{x}^2) \qquad (2.12)$$

But by (2.9), $n\bar{x} = x_1 + x_2 + \ldots + x_n$, hence

$$(x_1 - \bar{x})^2 + \ldots + (x_n - \bar{x})^2 = x_1^2 + x_2^2 + \ldots +$$
$$+ x_n^2 - 2n\bar{x}^2 + n\bar{x}^2 = x_1^2 + x_2^2 + \ldots + x_n^2 - n\bar{x}^2 \qquad (2.13)$$

Again, as for complete populations, any quantity A may be subtracted from each of $x_1, x_2, \ldots x_n$ and the 'coded' mean calculated and these coded values can be used either in formula (2.10) or (2.11) to calculate s.

Example

2.5(i). Calculate $\bar{x}$ and s^2 for the following two samples (observations of the percentage of ammonia in a plant gas) 37, 35, 43, 34, 36, 48; 37, 38, 36, 47, 48, 57, 38, 42, 49; (*a*) directly, (*b*) by subtracting 36 from each observation.

2.6 Qualitative classification of populations: frequency polygon, frequency curve; *U*-shaped and *J*-shaped populations, symmetrical and skew distributions; mode and median

This subsection can be omitted on a first reading; its subject matter is, in fact, of more value in economics and demography (the sampling of human populations) than in science or technology.

The data of a frequency diagram can often be shown more clearly by joining the tops of adjacent frequency ordinates by straight lines. The polygon thus produced is called a 'frequency' polygon and Fig. 2.3(a) shows the data of Fig. 2.1 presented this way.

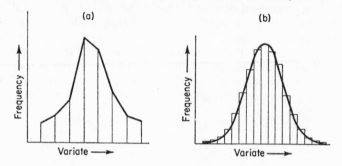

FIGURE 2.3. (*a*). Frequency polygon fitted to frequency diagram of Fig. 2.1. (*b*) Frequency curve fitted to histogram of Fig. 2.2

Where there are a large number of variate-values in the frequency polygon the polygon may be replaced by a smooth curve. Alternatively a curve may be drawn through the mid-points of the tops of the boxes of a histogram. Such a frequency curve for the data of Table 2.2 is shown in Fig. 2.3(b).

The shape of the frequency curve enables a qualitative classification of populations to be made. Fig. 2.4 shows some typical shapes of population ('distribution' is an alternative word for population). The descriptions are, in general, self-evident. However, the word 'mode' needs definition. It is the value of the variate for which the frequency is a *maximum* (i.e. a local maximum as in calculus). Some curves like the multi-modal curve of Fig. 2.4 have several modes.

Some of the distributions which have appeared earlier in this chapter belong to one or other of these population types. Thus the data of 'proportion of sky covered' of 2.1(i) is U-shaped for, in Greenwich as in many places, the sky is more often completely overcast or virtually cloudless than half-clouded. Again, the heights of men are

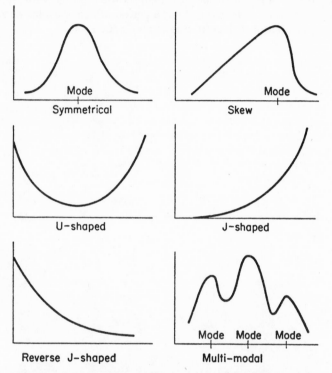

FIGURE 2.4. Various qualitative types of frequency curve

distributed practically 'symmetrically' (see Table 2.2) but the wage-rate table 2.1 is somewhat 'skew'.

An example of a reverse J-shaped distribution is the frequency of men with a given income as variate; a large number of men have low incomes, a moderate number have medium incomes and very few are millionaires.

Two further quantitative measures are sometimes calculated to describe more fully the nature of a population. First the moments μ_2, μ_3, μ_4 are computed (see 2.1), then the 'coefficients'

$$\beta_1 = \mu_3^2/\mu_2^3, \qquad \beta_2 = \mu_4/\mu_2^2 \tag{2.14}$$

β_1 measures the degree of 'skewness', (for a symmetrical distribution $\mu_3 = 0$ and therefore $\beta_1 = 0$). β_2 measures the 'kurtosis' or 'degree of peakedness'; a high value of β_2 means that the frequency curve is very peaked and a low value means that it is very broad.

One final term; the 'median' of a population or of a sample is that observation which has as many observations with smaller variate-values as with larger. After a period of near-obsolescence the median has had a revival in usage (see Chapter 14).

3

Probability, its meaning, real and theoretical populations

3.1 Probability for finite populations

By a 'finite' population we mean one with a finite (= countable) number of individuals. Thus, the population of wage-rates of Table 2.1 is finite, and so too are populations of men's heights, since the number of men in the world, though large, is countable. (It is possible to envisage *infinite* populations, as will be seen.)

Now, the reader may have some idea of what probability is, but, for the matters in hand, a precise and clear definition must be made. Since the aim of statistics is to gather as *much* information from as *few* observations as possible, consider the following question: If, from the wage-rate data of Table 2.1, an 'individual' is selected at random, what is the probability that this is 75 pence per hour (note that an 'individual' is an observation). The answer is 23/211, because 23 men out of 211 have this wage-rate and, if we repeated the operation of choosing a man at random a very large number of times, a 75 pence per hour man would be chosen a proportion 23/211 of times.

The perceptive reader will ask 'Exactly what does "at random" mean?' The answer is that there is no *pattern or regularity whatsoever* in the selection. (How to achieve random selection in practice is discussed in Chapter 15.)

We now express this idea in general terms. First we define probability precisely thus: The *probability* that a single observation has a given variate value, X say, is the *proportion* of times the variate-value X turns up when a very large number (theoretically, infinity) of random selections are made. We further ask the reader, as a reasonable person, to agree that, when selection is *random*, the proportion of times that a value X turns up *exactly equals* the *proportion* of individuals in the population who have a variate-value X, *in the long run*. (There is *no proof* of this; it is an *axiom*, that is, it is regarded as a self-evident truth.)

Hence, given a population consisting of f_1 individuals with variate-value X_1, of f_2 with variate-value X_2, and so on up to f_N with variate-value X_N, then the probability that a randomly selected individual has a variate-value X_i is

$$p_i = f_i/(f_1 + f_2 + \ldots + f_N) \qquad (3.1)$$

15

From this formula we can deduce the very important relation

$$
p_1 + p_2 + \ldots + p_N = \frac{f_1}{f_1 + f_2 + \ldots + f_N} + \frac{f_2}{f_1 + \ldots + f_N}
$$
$$
+ \ldots \cdot \frac{f_N}{f_1 + f_2 + \ldots + f_N} = \frac{f_1 + f_2 + \ldots + f_N}{f_1 + f_2 + \ldots + f_N} = 1 \quad (3.2)
$$

This can be put in words thus: The sum of the probabilities that an individual has one or other of the *possible* variate-values is *unity*.

Examples

3.1(i). Calculate the p_i (i.e. p_1 up to p_8) for the data of Table 2.1.

3.1(ii). Calculate the p_i for the data of Examples 2.1(i) (*a*), (*b*). What is the probability that the sky is half (0·5) or more clouded in 2.1(i) (*b*)?

Another relation is the following: Since the *proportion* of individuals with variate-value X_1 is the same as p_1 and so on, then the population mean μ can be found from the formula (see 2.1 and 3.1)

$$
\mu = p_1 X_1 + p_2 X_2 + \ldots + p_N X_N \quad (3.3)
$$

We now elaborate this notion of probability, occasionally using formulae which the reader may remember if he wishes. He should, however, make sure he understands the underlying ideas clearly.

3.2 Probability for infinite populations with continuous variate

The variate-values in many real populations (e.g. wage-rates) have only a finite number of possible values. But populations with a continuous range of possible variate-values can be envisaged, e.g. measurements of the velocity of light, for owing to variation from experiment to experiment even the same observer will never reproduce exactly the same value and, indeed, we are not *sure* that the velocity is an *absolute* constant.

The idea of a population with a continuous range has been so fruitful we shall consider it in some detail. It originated in astronomy when the observations made by observers at different places and/or by the same observer at different times had to be combined to give the best 'fix' for a given star. The chief cause of variation in the observations is atmospheric refraction which is varying all the time and the great mathematical physicist Gauss suggested that each observation should be regarded as the combination of the true position plus an 'error' term. This error term Gauss regarded as coming from a *population* of errors (with a continuous variate). By examining the variations of the observations *among themselves*, we can get quite a good idea of this error population, without a *detailed examination of the causes of the errors*. On plausible assumptions

Gauss worked out the properties of this error population and from these it is possible to deduce the overall best position for a given *star*, together with some idea of the final accuracy.

Later, the idea of continuous variate populations was extended, and the properties of real populations studied by considering continuous variate populations which *closely resemble* them. Thus, the 'heights of men' histogram of Fig. 2.1 is closely represented by the frequency curve (drawn through the midpoints of the tops of the

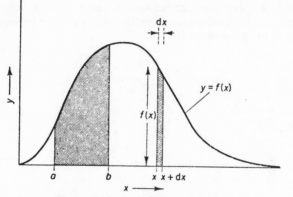

FIGURE 3.1. A frequency curve

boxes) of Fig. 2.3(b). In some respects this frequency curve is more realistic because it gives a better idea of how heights vary *inside* a box. But the great merit of such curves is that they may have a *simple formula* which enables properties to be calculated simply and quickly.

We consider now the extension of formula (3.1) for probability in finite populations to continuous variate populations. Since the frequency curve of a continuous population may have, in theory, any formula we shall call its equation $y = f(x)$, see Fig. 3.1. Instead of the frequency f_i of a variate-value X_i as in a frequency diagram or the frequency in a box of a histogram, we consider the frequency of variate-values in a narrow strip lying between x and $x + dx$, (see the shaded strip in Fig. 3.1). This frequency is

$$f(x)dx \tag{3.4}$$

Again the frequency of variate-values which lie anywhere between $x = a$ and $x = b$ is the integral (see the stippled area of Fig. 3.1)

$$\int_{x=b}^{x=a} f(x)dx \tag{3.5}$$

To specify completely the population we need to know the lowest possible variate-value (l, say) and the greatest value (g, say). Then the total frequency (i.e. the total number of individuals) in the population is

$$\int_{x=l}^{g} f(x)\mathrm{d}x \tag{3.6}$$

Analogous to (3.1), the *probability* that an individual selected at random has a variate-value between x and $x + \mathrm{d}x$ is equal to the *proportion* of the total frequency with a variate-value between x and $x + \mathrm{d}x$. This proportion is

$$\frac{f(x)\mathrm{d}x}{\int_{l}^{g} f(x)\mathrm{d}x} \tag{3.7}$$

The quantity

$$p(x) \equiv \frac{f(x)}{\int_{l}^{g} f(x)\mathrm{d}x} \tag{3.8}$$

is called the 'probability density function' of the population. It is analogous to the p_i of Section 3.1 and its definition should be remembered. Again, analogous to the important relation (3.2), we have the formula

$$\int_{x=l}^{g} p(x)\,\mathrm{d}x = \frac{\int_{x=l}^{g} f(x)\,\mathrm{d}x}{\int_{x=l}^{g} f(x)\,\mathrm{d}x} = 1 \tag{3.9}$$

or, in words, the integral of the probability density function $p(x)$ over all possible values of the variate (i.e. from $x = l$ to $x = g$) is unity.

(The reader must not think that *all* theoretical populations have continuous variates. In some problems we consider discrete objects such as particles, or customers, arriving at random. The number of customers is often the variate considered and this can only be 0, 1, 2, . . . and, for example, cannot be 3·279.)

The term 'distribution' is nowadays very commonly used instead of 'population' especially for theoretical populations.

We finish this section with the formula for the probability that an individual, selected at random, has a variate-value lying somewhere between a and b; this can be seen, from (3.5) and (3.8), to be

$$\int_{x=a}^{b} p(x)\mathrm{d}x \qquad (3.10)$$

a result which should be remembered.

Examples

3.2(i). If a frequency curve has formula $f(x) = \mathrm{e}^{-x}$, and $l = 0$, $g = \infty$, show that $p(x) = \mathrm{e}^{-x}$. If $f(x) = \mathrm{e}^{-\alpha x}$, $l = 0$, $g = \infty$, show that $p(x) = \alpha \mathrm{e}^{-\alpha x}$ (both these are called 'negative exponential' distributions).

3.2(ii). If $f(x) = x^2 - 2x + 5$, $l = 0$, $g = 3$, find the probability density function $p(x)$. Find the probability that an observation made at random lies between 1 and 2.

3.3 Sampling populations (distributions)

These are theoretical populations which arise thus: Suppose we take a sample of a given size from a population and calculate the sample mean $\bar{x}$ from formula (2.9). If we did this a very large number of times we could plot the curve of frequency of occurrence of different values of $\bar{x}$. In other words we would get a population of sample means. Such a population is called a 'sampling population' or 'sampling distribution'. If we repeated this process for samples of all sizes we can find out various vitally useful properties.

Consider, for example the fact that the sample mean is used to estimate the true or population mean. The sample mean though usually close to the population mean does differ from it. The sampling population properties enable us to say what size of sample is required to estimate the population mean with a given accuracy.

We can, each time we take a sample, calculate the sample variance s^2, (from formula (2.11)) and form a population based on the frequency of occurrence of different values of s^2. Or we could measure the sample 'range' (= largest − smallest observation) and obtain a population from the values of this. In fact any population derived from an original population by taking samples repeatedly and calculating some function of the observations in the sample is called a 'sampling population'.

If the original population has a reasonably simple mathematical formula, the formulae for the sampling distributions derived from it can often (though not always) be found (though sometimes some rather clever mathematical tricks have to be used). In fact these mathematical formulae form the basis of modern statistics and its very powerful methods.

There is no need to follow the mathematics in detail to appreciate these methods, however, and they are, in general, merely outlined here. However, since the useful formulae are usually given in terms of probability rather than frequency we now rephrase a few definitions and define some useful technical terms.

3.4 Moments of populations in terms of probability

Section 2.4 gives the formulae for the moments of a finite population. In terms of the probabilities p_i (see formula (3.1)) it will be seen that

$$\nu_r = p_1 X_1^r + p_2 X_2^r + \ldots + p_N X_N^r \tag{3.11}$$

where ν_r is the rth moment about the origin. The corresponding formula for continuous variate populations is

$$\nu_r = \int_{x=l}^{g} p(x)\, x^r \mathrm{d}x \tag{3.12}$$

where $p(x)$ is the probability density function (see formula (3.8)).
Similarly the rth moment about the mean μ_r has the formula

$$\mu_r = p_1(X_1 - \mu)^r + p_2(X_2 - \mu)^r + \ldots + p_N(X_N - \mu)^r \tag{3.13}$$

or

$$\mu_r = \int_{x=l}^{g} (x - \mu)^r p(x)\mathrm{d}x \tag{3.14}$$

where μ is the population mean, i.e. $\mu = \nu_1$.
Note the following relations

$$\nu_0 = \mu_0 = \int_{x=l}^{g} p(x)\mathrm{d}x = 1 \tag{3.15}$$

which is the same as formula (3.9), and

$$\mu_1 = \int_{l}^{g} (x - \mu)p(x)\mathrm{d}x = \int_{l}^{g} xp(x)\mathrm{d}x - \mu \int_{l}^{g} p(x)\mathrm{d}x$$
$$= \mu - \mu = 0 \tag{3.16}$$

Note that μ_2 is identical with the variance σ^2 just as for finite populations (see Section 2.4), for it is the mean value of the square of the deviations of each observation (i.e. it is the mean value of $(x - \mu)^2$).

Examples

3.4(i). Find ν_1, ν_2, μ_1, μ_2 for the population with $p(x) = \mathrm{e}^{-x}$, $l = 0$, $g = \infty$.

3.4(ii). For the population with $p(x) = \frac{1}{3}$, $l = 1$, $g = 4$ find ν_1, ν_2, ν_r; μ_1, μ_2, μ_r. (This is called a 'rectangular' distribution, see Section 10.5.)

There are relations between the ν_r and the μ_r identical with those of (2.7) and (2.8) for finite populations, namely

$$\nu_1 = \mu; \quad \nu_2 = \mu_2 + \mu^2 = \sigma^2 + \mu^2 \tag{3.17}$$

$$\mu_3 = \nu_3 - 3\nu_2\nu_1 + 2\nu_1^3; \quad \mu_4 = \nu_4 - 4\nu_3\nu_1 + 6\nu_2\nu_1^2 - 3\nu_1^4 \tag{3.18}$$

Example

3.4(iii). Prove (3.17) and (3.18) (for the mathematically curious).

3.5 The 'expected' value

This is a simple but very useful notation. Suppose we select at random a sample from a population and calculate some quantity such as the mean or variance or 'range' (see Section 3.3). Then the *mean value* of this quantity averaged over a large number (theoretically, infinity) of randomly selected samples (of the same size) is called the 'expected value' of this quantity (and denoted by the symbol E). Thus, if x denotes the variate-value of a *single* individual selected at random (i.e. a sample of size 1), then

$$E(x) = \mu \tag{3.19}$$

because the mean value of x averaged over a large number of selections is equal to the population mean μ (since we agree that, in the long run, each possible variate-value will appear its correct proportion of times). In the case of a finite population with possible variate-values $X_1, X_2, \ldots X_N$ whose probabilities are $p_1, p_2, \ldots p_N$ then

$$E(x) = p_1 X_1 + p_2 X_2 + \ldots + p_N X_N = \mu \tag{3.20}$$

We can express the moments about the origin (ν_r) and those about the mean (μ_r) conveniently in this notation, thus

$$E(x^r) = p_1 X_1^r + p_2 X_2^r + \ldots + p_N X_N^r = \nu_r;$$
$$E(x - \mu)^r = p_1(X_1 - \mu)^r + \ldots + p_N(X_N - \mu)^r = \mu_r \tag{3.21}$$

Similarly, for continuous variates,

$$E(x^r) = \int_l^g x^r p(x) \mathrm{d}x = \nu_r; \quad E(x - \mu)^r = \int_l^g (x - \mu)^r p(x) \mathrm{d}x = \mu_r \tag{3.22}$$

and, in particular,

$$E(x) = \int_l^g xp(x)\mathrm{d}x = \mu$$

$$E(x - \mu)^2 = \int_l^g (x - \mu)^2 p(x)\mathrm{d}x = \mu_2 = \sigma^2 \tag{3.23}$$

3.6 The cumulative distribution function (for continuous variates)

The integral of the probability density $p(x)$ from the lowest value l of x up to a given value $x = X$, that is

$$\int_{x=l}^{X} p(x)\mathrm{d}x = P(X), \text{ say} \tag{3.24}$$

is called the '(cumulative) distribution function'; the word 'cumulative' is often omitted. Note that X may take any value between l and g.

We shall not use the distribution function explicitly in this book, but it is often very useful in the theory. Note that the derivative of the distribution function equals the probability density function, that is

$$\mathrm{d}P(X)/\mathrm{d}X = p(X) \tag{3.25}$$

4

Basic properties of the normal distribution

4.1 Origin and pre-eminence of the normal distribution

The normal distribution is the continuous variate distribution which Gauss derived for the errors in making astronomical observations, as mentioned in Section 3.2. He considered that the *total error* in making an observation is the sum of a very large number of very small errors each of which might be *positive* or *negative* at *random*. He then proved mathematically that, if x is the total error, the population of total errors has a probability density function (defined in Section 3.2) of the form

$$p(x) = \frac{1}{k\sqrt{(2\pi)}} \exp\left(-\frac{x^2}{2k^2}\right) \tag{4.1}$$

where k is large if the total errors are large, and small if they are small. Gauss showed further that k, in fact, is equal to the standard deviation σ of this population; and, nowadays, we write directly

$$p(x) = \frac{1}{\sigma\sqrt{(2\pi)}} \exp\left(-\frac{x^2}{2\sigma^2}\right) \tag{4.2}$$

The reader should note that no attempt need be made to ascertain the *nature* of the *individual causes* of the small errors, their *overall effect* can be ascertained from the total error population *itself*. This basic idea has been extended to practically every type of observation or measurement in science and technology and elsewhere. For, in many cases, it is too costly or the outcome too uncertain to investigate these sources of small errors but by using Gauss's normal distribution we can, nevertheless, assess their overall effect. In fact, wherever there are a number of sources of small errors we can safely assume that the normal distribution applies, thus dealing with one of the basic problems of statistics namely how to classify the data.

Further, this distribution has been applied to a different type of observation, namely biological. For example, consider the heights of men. The height of a man selected at random can be considered as composed of a standard value, the average man's height, plus a variation term (equivalent to the total error) which is the sum of the small variations due to a large number of factors such as the genes of race, of parents, of grandparents, etc., and of many environmental

23

factors. Practical measurements of men's heights have shown that they have a distribution very close to the normal and so too have many other biological measurements.

In addition, certain features of non-normal populations have been shown to have close affinities to the normal (e.g. the sampling distribution of the means of samples). So universal is this type of distribution that its old name of error curve has been replaced by 'the normal distribution'.

4.2 Formulae for the general normal distribution

The normal population with probability density given by (4.2) has a mean value of zero. However, in general a normal distribution will have a non-zero true mean (which we denote, as usual by μ), and denoting its standard deviation by σ, as usual, its probability density function has the formula

$$p(x) = \frac{1}{\sigma\sqrt{(2\pi)}} \exp\left[-\frac{(x-\mu)^2}{2\sigma^2} \right] \qquad (4.3)$$

(For the heights data of Table 2.2, $\mu = 67\cdot46$ in. and $\sigma = 2\cdot57$ in.)

To complete the general normal distribution we need the least and greatest possible variate values l and g. Theoretically these are $l = -\infty$, and $g = +\infty$, *whatever* the value of μ. It is often the case that it is not *physically* possible for variate-values of $+$ or $-\infty$ to exist (e.g. heights of men). However, as will be shown shortly, the probabilities associated with large positive or negative deviations from the mean μ are so very small that this discrepancy rarely has any practical effect.

It can be proved mathematically that

$$p(x) = \frac{1}{\sigma\sqrt{(2\pi)}} \int_{-\infty}^{+\infty} \exp\left[-\frac{(x-\mu)^2}{2\sigma^2} \right] dx = 1 \qquad (4.4)$$

a result to be expected from the general formula (3.9).

If the values of μ and σ are both known, then the probability density $p(x)$ of (4.3) can be calculated for each value of x and a frequency curve of $p(x)$ can be plotted (as in Figs. 4.1 and 4.2), thus completely specifying the distribution.

4.3 Probabilities associated with the normal distribution

Fig. 4.1 shows the curve formed by the probability density of the general normal distribution. Note that the curve is symmetrical about the true mean μ and that the ordinate (in this case $\sigma\, p(x)$) becomes very small indeed for values of the variate differing from μ by more than 3σ. Note also that a change in μ merely shifts the

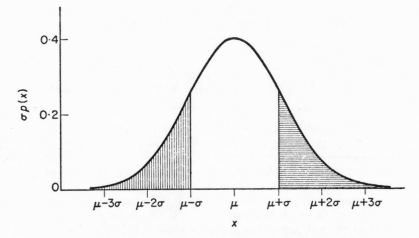

FIGURE 4.1. Frequency curve for the general normal distribution
μ, σ are the population mean, standard deviation; $p(x)$ is the
probability density; x is the variate

curve to left or right but a change in σ alters its appearance somewhat. Fig. 4.2 shows two normal distributions, the first with $\mu = 2$
and $\sigma = 2/3$, and the second with $\mu = 5$ and $\sigma = 1.5$.

Some idea of the rapid fall off in probability density with increase
in deviation from the mean is given by Table 4.1. A fuller table of

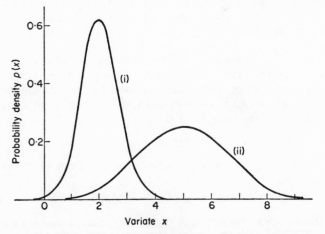

FIGURE 4.2. Two normal distribution frequency curves:
(i) $\mu = 2$, $\sigma = 2/3$; (ii) $\mu = 5$, $\sigma = 1.5$

Table 4.1

Some values of $\sigma p(x)$, where $p(x) = $ normal probability density function

$x - \mu$	$\pm 4 \cdot 5\sigma$	$\pm 4 \cdot 0\sigma$	$\pm 3 \cdot 0\sigma$	$\pm 2 \cdot 0\sigma$	$\pm 1 \cdot 0\sigma$	± 0
$\sigma p(x)$	0·00002	0·00013	0·00443	0·05399	0·24197	0·39894

probability densities is given in the Appendix, Table A1. It can be seen from these tables that the probabilities associated with large deviations from the mean are very small indeed.

More practically useful than the probability density is the probability that an individual selected at random has a variate-value *greater* than some given value. For example, what is the probability that a man selected at random is over 6 ft 2 in. in height (this equals the proportion of men in the population over 6 ft 2 in. tall, and such considerations have to be taken into account in, say, fixing the height of ceilings in buses).

The proportion of a normal population with variate-values greater than, say, $\mu + \sigma$, is, applying formula (3.10) with $a = \mu + \sigma$ and $b = +\infty$,

$$\int_a^b p(x)\mathrm{d}x = \int_{x=\mu+\sigma}^\infty \frac{1}{\sigma\sqrt{(2\pi)}} \exp\left[-\frac{(x-\mu)^2}{2\sigma^2}\right]\mathrm{d}x = 0{\cdot}15866 \quad (4.5)$$

(Note that such integrals have to be calculated numerically.) This proportion is shown shaded horizontally in Fig. 4.1; such an area going from a given variate-value to infinity is called a 'tail'. Since the normal distribution is symmetrical about its mean $x = \mu$, the proportion which has variate-values *less* than $\mu - \sigma$ (shown shaded vertically in Fig. 4.1) is also 0·15866. This area is called a 'left-hand tail' and the horizontally shaded area a 'right-hand tail'.

We now give a very convenient notation for the probability that an individual selected at random has a variate-value x greater than, or equal to, a given value k; it is

$$\Pr(x \geqslant k) \quad (4.6)$$

If $x \geqslant \mu + \sigma$ then, of course $x - \mu \geqslant \sigma$; so we can write formula (4.5) as

$$\Pr(x - \mu \geqslant \sigma) = 0{\cdot}15866 \quad (4.7)$$

(For continuous variates the probability that $x = k$ exactly is infinitesimal, so that $\Pr(x \geqslant k) = \Pr(x > k)$; for discrete variates, this is not true.) Table 4.2 shows some probabilities associated with 'tails' of the normal distribution. A fuller table is in the Appendix, Table A2. Note that these probabilities fall off with increase of deviation $x - \mu$ even faster than the probability density, see Table 4.2.

Table 4.2. PROBABILITIES ASSOCIATED WITH TAILS OF THE NORMAL DISTRIBUTION

	Right-hand	Left-hand	
$\Pr(x - \mu \geqslant 0)$	$= \Pr(x - \mu \leqslant 0)$		$= 0\cdot50000$
$\Pr(x - \mu \geqslant \sigma)$	$= \Pr(x - \mu \leqslant - \sigma)$		$= 0\cdot15866$
$\Pr(x - \mu \geqslant 2\sigma)$	$= \Pr(x - \mu \leqslant - 2\sigma)$		$= 0\cdot02275$
$\Pr(x - \mu \geqslant 3\sigma)$	$= \Pr(x - \mu \leqslant - 3\sigma)$		$= 0\cdot00135$
$\Pr(x - \mu \geqslant 4\sigma)$	$= \Pr(x - \mu \leqslant - 4\sigma)$		$= 0\cdot00003$

For certain tests (see Chapter 6) we need the probability that the variate-value x of a randomly selected individual *differs* from μ (i.e. *either* x is greater than, or x is smaller than, μ by a given amount). Thus, the probability that x differs from μ by 2σ or more is the sum of $\Pr(x - \mu \geqslant +2\sigma)$ and of $\Pr(x - \mu \leqslant -2\sigma)$. This combined result is written

$$\Pr(|x - \mu| \geqslant 2\sigma) = 0\cdot02275 + 0\cdot02275 = 0\cdot0455$$

(N.B. If y is any number, positive or negative, $|y|$ denotes the absolute magnitude of y, e.g. if $y = -6$, $|y| = +6$; if $y = 4$, $|y| = +4$; $|y|$ is *always* positive and is called the 'modulus' of y.) Such probabilities are called 'two-tailed', whereas those of Table 4.2 are 'one-tailed'.

The most useful probabilities, in practice, are those given in Table 4.3; they are of the greatest importance in the 'significance' tests of Chapter 6 and are often called 'critical probabilities', for reasons given in Section 6.2.

Table 4.3. 'CRITICAL' PROBABILITIES OF THE NORMAL DISTRIBUTION

Two-tailed	One-tailed
$\Pr(|x-\mu| \geqslant 1\cdot96\sigma) = 0\cdot0500 = 5\% = \Pr(x-\mu \geqslant 1\cdot645\sigma) = \Pr(x-\mu \leqslant -1\cdot645\sigma)$	
$\Pr(|x-\mu| \geqslant 2\cdot58\sigma) = 0\cdot0100 = 1\% = \Pr(x-\mu \geqslant 2\cdot33\sigma) = \Pr(x-\mu \leqslant -2\cdot33\sigma)$	
$\Pr(|x-\mu| \geqslant 3\cdot29\sigma) = 0\cdot0010 = 0\cdot1\% = \Pr(x-\mu \geqslant 3\cdot09\sigma) = \Pr(x-\mu \leqslant -3\cdot09\sigma)$	

Examples

4.3(i). Using Table 4.1 plot frequency curves for the normal distribution with (a) $\mu = 4$, $\sigma = 2$; (b) $\mu = -2$, $\sigma = 0\cdot8$.

4.3(ii). If, for heights of Englishmen, $\mu = 5$ ft 8 in., $\sigma = 2\cdot5$ in., find from Table A2 what proportion of Englishmen are (a) over 6 ft 3 in., (b) under 5 ft.

4.3(iii). Taking $\mu = 5$ ft 8 in., $\sigma = 2\cdot5$ in., find the height above which there are $2\cdot5\%$ of Englishmen, and that below which there are 10%.

4.4 Moments of the normal distribution

The rth moment about the origin, ν_r, is, by its definition (see (3.12))

$$\nu_r = \int_{x=-\infty}^{\infty} x^r \left[\frac{1}{\sigma\sqrt{(2\pi)}}\right] \exp\left[-\frac{(x-\mu)^2}{2\sigma^2}\right] dx \qquad (4.8)$$

Such integrals can be evaluated exactly (unlike that of 4.5), giving

$$\nu_0 = 1, \quad \nu_1 = \mu, \quad \nu_2 = \sigma^2 + \mu^2,$$
$$\nu_3 = 3\mu\sigma^2 + \mu^3, \quad \nu_4 = 3\sigma^4 + 6\mu^2\sigma^2 + \mu^4 \qquad (4.9)$$

The rth moment about the mean, μ_r, is, by definition (see (3.14))

$$\mu_r = \int_{x=-\infty}^{\infty} (x-\mu)^r \left[\frac{1}{\sigma\sqrt{(2\pi)}}\right] \exp\left[-\frac{(x-\mu)^2}{2\sigma^2}\right] dx \quad (4.10)$$

These integrals can also be evaluated exactly and we have

$$\mu_0 = 1, \quad \mu_1 = 0, \quad \mu_2 = \sigma^2, \quad \mu_3 = 0, \quad \mu_4 = 3\sigma^4, \text{ etc.} \quad (4.11)$$

Example

4.4(i). Find $\nu_2, \nu_3, \nu_4, \mu_3, \mu_4$ for the cases (*a*) $\mu = 4$, $\sigma = 2$; (*b*) $\mu = -2$, $\sigma = 0\cdot8$; (*c*) $\mu = 68$ in., $\sigma = 2\cdot5$ in.

4.5 Distribution of the sum, of the difference, of two (or more) observations from separate normal populations

If x is a random observation from a normal population with mean μ_1 and variance $\sigma_1{}^2$, and y is a random observation from another normal distribution with mean μ_2 and variance $\sigma_2{}^2$, then it can be proved that the sum of the two observations, i.e. $x + y$, is distributed normally with mean $\mu_1 + \mu_2$ and with variance $\sigma_1{}^2 + \sigma_2{}^2$.

The difference $x - y$ is also distributed normally with mean $\mu_1 - \mu_2$, but with the *same* variance $\sigma_1{}^2 + \sigma_2{}^2$.

Further, if $x, x', x'', \ldots$ are random observations from several normal distributions with means $\mu, \mu', \mu'', \ldots$ and standard deviations $\sigma, \sigma', \sigma'', \ldots$, then $x \pm x' \pm x'' \pm \ldots$ is normally distributed with mean $\mu \pm \mu' \pm \mu'' \pm \ldots$, provided the sequence of signs is the *same*, but with variance $\sigma^2 + (\sigma')^2 + (\sigma'')^2 + \ldots$ *whatever* the signs.

Examples

4.5(i). Marksmen A and B have average scores of $473\cdot2$, $481\cdot4$ (out of 500) with standard deviations $2\cdot4$ and $3\cdot2$ respectively, what is (a) the mean and (b) the standard deviation of the sum of their scores? (Assume their scores are distributed normally.)

4.5(ii). Find the mean and standard deviation of the difference of the scores given in 4.5(i).

4.5(iii). Given that x, y, z are normally distributed variates with means 11·3, 12·8, 9·7 and standard deviations 1·3, 0·7, 2·1 find the mean and standard deviation of (a) $x + y + z$, (b) $x + y - z$, (c) $x - y - z$.

5

Some properties of sampling distributions

The results of this chapter are *most important*; the proofs can be remembered if so desired, but are not essential.

5.1 The importance of sampling distributions; (definition of a 'statistic')

Suppose that $x_1, x_2, \ldots x_n$ are the variate-values of the individuals in a sample of size n. Then the sample mean $\bar{x}$ is

$$\bar{x} = (x_1 + x_2 + \ldots + x_n)/n$$

If we repeatedly take samples of size n we can get a 'population' of different values of $\bar{x}$ called a sampling distribution (first defined in Section 3.3), and it is possible to work out, in theory, the properties of such a distribution. (Note that two different samples can give the *same* value of $\bar{x}$, e.g. 2, 4, 5, 9; and 1, 3, 6, 10 both have a mean 5, so that finding the probability that $\bar{x}$ has a given value may not be a simple matter.)

The mean is not the only quantity of interest and, for example, the sum of squares $x_1^2 + x_2^2 + \ldots + x_n^2$ is of great importance. Nowadays a general term has been introduced and any function of $x_1, x_2, \ldots x_n$ is called a 'statistic' (this word is always used in the singular to avoid confusion with the subject statistics).

The properties of the sampling distribution of a given statistic such as the mean, the sum of squares, the range (= largest x − smallest x) etc., form the basis of statistics. We shall now discuss a few of these properties.

5.2 The expected value of the mean of a sample = Population mean (or $E(\bar{x}) = \mu$)

The expected value (see Section 3.5) is the arithmetical mean value average over an infinite number of selections. Before finding the expected value of the mean of a sample we shall consider two simple theorems. If x denotes the variate-value of an individual what is the expected value of kx, where k is a constant. Now we know that $E(x) = \mu$, the population mean (see Section 3.5); hence, if each time a selection is made x is multiplied by k, the average value of the product kx must be $k\mu$; that is

$$E(kx) = k\mu \qquad (5.1)$$

a simple but important result. Again, if x and y are the variate-values of two individuals chosen from two separate populations, then it can be seen that

$$E(x + y) = E(x) + E(y) \qquad (5.2)$$

$$E(x - y) = E(x) - E(y) \qquad (5.3)$$

(As an example consider two marksmen, one who averages 97 out of 100 and the other 95 out of 100; in a large number of shoots their combined score will average $97 + 95 = 192$; whether they both do well at the same time or not.) The two populations from which x and y are drawn may be different or may be the same, but in the latter case both x and y must be chosen *completely* at random, e.g. x could not be chosen at random and then y from among those individuals *smaller* than x.

An individual selected at random from a population is sometimes called a random variable. Theorem (5.2) can be extended to the sum of any number of individuals from different populations (i.e. to the sum of any number of 'random variables'). Thus, if a, b, c, . . . h are 'random variables' from different, or from the *same*, populations, then

$$E(a + b + c + . . . + h) = E(a) + E(b) + . . . + E(h) \qquad (5.4)$$

Since $E(x) = \mu$ where x is an individual from a given population (with true mean μ) applying (5.4) to $x_1, x_2, . . . x_n$, when $x_1, . . . x_n$ are n individuals selected at random from this population, then

$$E(x_1 + x_2 + . . . + x_n) = E(x_1) + E(x_2) + . . . + E(x_n)$$
$$= \mu + \mu + . . . + \mu = n\mu \qquad (5.5)$$

We now extend 5.1 to 5.5 thus obtaining

$$E[k(x_1 + . . . + x_n] = E(kx_1 + . . . + kx_n)$$
$$= k\mu + k\mu + . . . + k\mu = k(n\mu)$$

Now put $k = 1/n$ and, since $(1/n)(x_1 + x_2 + . . . + x_n) = \bar{x}$, we have

$$E\{(1/n)(x_1 + . . . + x_n)\} = E(\bar{x}) = (1/n)(n\mu) = \mu \qquad (5.6)$$

That is 'the expected value of the mean of a sample (whatever its size n) is μ, the population mean'; a most *important result*. Note that this is true whatever the type of population (e.g. it may be a normal, or non-normal, finite, or infinite population and may have a continuous variate or not, etc.).

5.3 The expected value of the product of two independent random variables = the product of their expected values

Let x and y be two random variables. Then provided x and y are selected *independently* of each other

$$E(xy) = E(x)E(y) \qquad (5.7)$$

another *important result*. The proof is as follows; suppose that x is selected from the finite population whose possible variate-values are $X_1, X_2, \ldots X_M$ with probabilities $p_1, p_2, \ldots p_M$, and that y is selected from the population with variate-values $Y_1, Y_2, \ldots Y_N$ and probabilities $p_1', p_2', \ldots p_N'$. Now in a very large number of selections the variate-value X_1 will be picked a proportion p_1 of times and associated with it will be Y_1 or Y_2 or $\ldots$ or Y_N, each occurring its correct proportion of times; similarly for X_2, etc. Hence

$$
\begin{aligned}
E(xy) &= p_1 X_1(p_1' Y_1 + p_2' Y_2 + \ldots + p_N' Y_N) \\
&\quad + \ldots + p_M X_M(p_1' X_1 + \ldots + p_N' Y_N) \\
&= (p_1 X_1 + \ldots + p_M X_M)(p_1' Y_1 + \ldots + p_N' Y_N) \\
&= E(x)E(y) \qquad (5.8)
\end{aligned}
$$

since $p_1 X_1 + \ldots + p_M X_M = E(x)$, and $p_1' Y_1 + \ldots + p_N' Y_N = E(y)$, see (3.20). The proof can be extended to continuous variate populations quite simply.

(Let us see what happens if x and y are not selected independently. Consider the simple populations $X_1, X_2 = 1, 5$; $p_1, p_2 = 0.4, 0.6$; and $Y_1, Y_2 = 2, 8$; $p_1', p_2' = 0.3, 0.7$. If x and y are selected independently of each other then

$$
\begin{aligned}
E(xy) &= E(x)E(y) \\
&= (0.4 \times 1 + 0.6 \times 5)(0.3 \times 2 + 0.7 \times 8) = 21.08
\end{aligned}
$$

If however, X_1 and X_2 are selected at random but Y_1 is always selected with X_1, and Y_2 with X_2, and the pairs X_1, Y_2; X_2, Y_1 are never chosen

$$E(xy) = 0.4 \times 1 \times 2 + 0.6 \times 5 \times 8 = 24.8$$

which is larger than 21.08; non-random selection usually gives a larger value.)

Example

5.3(i). Calculate $E(x)$, $E(y)$, $E(x^2)$, $E(xy)$, $E(y^2)$, $E(x^2 y^2)$ for the populations $X_1, X_2, X_3 = 0, 3, 4$; $p_1, p_2, p_3 = 0.2, 0.3, 0.5$; $Y_1, Y_2, Y_3 = -1, 1, 2$; $p_1', p_2', p_3' = 0.1, 0.3, 0.6$; assuming independent selection. (Note, $E(x^2) = p_1 X_1^2 + p_2 X_2^2 + p_3 X_3^2$.)

5.4 The variance of the sum or difference of several independent random variables = the sum of their variances

Now, as has been stated earlier, the variance of a population is the expected value of the square of the difference between the variate-value x and the true mean μ. This result can be written, since $E(x) = \mu$, see (3.19),

$$\text{var}(x) = E(x - \mu)^2 = E\{x - E(x)\}^2 \qquad (5.9)$$

Now applying (5.9) to the variance of the sum of x and y we have

$$\text{var}(x + y) = E\{x + y - E(x + y)\}^2 \qquad (5.10)$$

If $E(x) = \mu_1$, $E(y) = \mu_2$, say, then the right-hand side of (5.10) becomes

$$E\{x + y - \mu_1 - \mu_2\}^2 =$$
$$E\{(x - \mu_1)^2 - 2(x - \mu_1)(y - \mu_2) + (y - \mu_2)^2\} =$$
$$E(x - \mu_1)^2 - 2E\{(x - \mu_1)(y - \mu_2)\} + E(y - \mu_2)^2 \qquad (5.11)$$

Now, since $x - \mu_1$ and $y - \mu_2$ are independently selected, by (5.1)

$$E\{(x - \mu_1)(y - \mu_2)\} = E(x - \mu_1)E(y - \mu_2)$$

But $E(x - \mu_1) = E(x) - E(\mu_1) = \mu_1 - \mu_1 = 0$; again, $E(x - \mu_1)^2$ $= \text{var}(x)$ and $E(y - \mu_2)^2 = \text{var}(y)$, hence (5.11) becomes

$$\text{var}(x + y) = E(x - \mu_1)^2 + E(y - \mu_2)^2$$
$$= \text{var}(x) + \text{var}(y) \qquad (5.12)$$

This result can be extended to any number of independent random variables, and if $a, b, c, \ldots h$ are such variables then

$$\text{var}(a + b + c + \ldots + h) = \text{var}(a) + \text{var}(b)$$
$$+ \ldots + \text{var}(h) \qquad (5.13)$$

another *most important result*. In words 'The variance of the sum of several independent random variables is the sum of their separate variances'. If the variables are *dependent* in some way on each other the variance is usually *larger*.

Example

5.4(i). Find the variance, assuming independence, of $x + y$ for the data of Example 5.3(i).

It can also be proved that

$$\text{var}(x - y) = \text{var}(-x + y) = \text{var}(x) + \text{var}(y) \qquad (5.14)$$

and, more generally still,

$$\text{var}\,(\pm a \pm b \pm c \pm \ldots \pm h) = \text{var}\,(a)$$
$$+ \text{var}\,(b) + \ldots + \text{var}\,(h) \quad (5.15)$$

whatever the combination of $+$ and $-$ signs on the left-hand side. So, whether we consider the sum or difference, or part sum and part difference of several independent random variables, its variance is the *sum* of the separate variances.

Example

5.4(ii). Find var $(x - y)$ for the data of Example 5.3(i).

5.5 The variance of $\bar{x}$ the sample mean $= \sigma^2/n$

Since the true mean of a population is usually estimated from a sample mean the variance of the latter is important in assessing accuracy. We need three simple preliminary results. We have seen that $E(kx) = kE(x)$, see (5.1); similarly

$$E(kx)^2 = E(k^2x^2) = k^2E(x^2) \quad (5.16)$$

Our second result follows from this; it is

$$\text{var}\,(kx) = E\{kx - E(kx)\}^2 = E(kx - k\mu)^2$$
$$= E\{k^2(x - \mu)^2\} = k^2\,\text{var}\,(x) \quad (5.17)$$

Thirdly, if x_1 and x_2 are independent observations from the same population

$$\text{var}\,(x_1) = \text{var}\,(x_2) \quad (5.18)$$

(This last result occasionally presents difficulty to students who think that because x_2 is a different observation from x_1, it must have different properties; in the *long run*, however, the *same set* of possible variate-values will appear the *same proportion* of times for *both*.)

If $x_1, x_2, \ldots x_n$ are the variate-values of a sample, each being selected independently of the others, then by (5.13),

$$\text{var}\,(x_1 + x_2 + \ldots + x_n) = \text{var}\,(x_1)$$
$$+ \text{var}\,(x_2) + \ldots + \text{var}\,(x_n) \quad (5.19)$$

If σ^2 is the population variance then-

$$\text{var}\,(x_1) = \text{var}\,(x_2) = \ldots = \text{var}\,(x_n) = \sigma^2 \quad (5.20)$$

Hence, from (5.19) it follows that

$$\text{var}\,(x_1 + x_2 + \ldots + x_n) = n\sigma^2 \quad (5.21)$$

Applying (5.17) with $k = 1/n$, remembering that $(1/n)(x_1 + \ldots + x_n) = \bar{x}$

$$\text{var}(\bar{x}) = \text{var}\{(1/n)(x_1 + \ldots + x_n)\}$$
$$= (1/n^2)(n\sigma^2) = \sigma^2/n \quad (5.22)$$

This important result is often expressed in a slightly different manner, since $\bar{x}$ is used as an estimate of μ, the difference $\bar{x} - \mu$ can be regarded as the 'error' in using $\bar{x}$. Now, by definition of variance, and, since $E(\bar{x}) = \mu$

$$\text{var}(\bar{x}) = E(\bar{x} - \mu)^2 = \sigma^2/n \quad (5.23)$$

The expected value of this 'error' squared is thus σ^2/n. Hence the 'standard' (i.e. root-mean-square) value of the error is

$$\sigma/\sqrt{n} \quad (5.24)$$

which is often expressed thus 'The standard error of the mean (of a sample) is $\sigma/\sqrt{n}$'.

Example

5.5(i). For the population of Example 4.3(ii) find the variance of the mean of a sample of size 25, 64, 79. What value of n would give a standard error of the mean of 0·25 in., of 0·05 in.?

5.6 $\bar{x}$ and s^2 are 'unbiased' estimates of μ and σ^2

As mentioned in Section 2.5, μ and σ^2 are estimated by calculating the mean $\bar{x}$ and observed variance $s^2 = \{(x_1 - \bar{x})^2 + \ldots + (x_n - \bar{x})^2\}/(n - 1)$, see formula (2.10), from the observations in a sample. Now the expected or mean value of $\bar{x}$ averaged over a large number of samples (of the same size, of course) is μ itself (as shown in Section 5.2, $E(\bar{x}) = \mu$). For this reason $\bar{x}$ is called an 'unbiased' estimate of μ. We shall now show that s^2 is an unbiased estimate of σ^2, i.e. we shall show that $E(s^2) = \sigma^2$.

We need two simple preliminary results. If x is a single observation

$$E(x^2) = E(x - \mu + \mu)^2 = E(x - 0)^2$$
$$+ E\{2\mu(x - \mu)\} + E(\mu^2) \quad (5.25)$$

Since $E(x - \mu)^2 = \text{var}(x) = \sigma^2$, $E(x - \mu) = E(x) - \mu = \mu - \mu = 0$, and $E(\mu^2) = \mu^2$, then

$$E(x^2) = \sigma^2 + \mu^2 \quad (5.26)$$

Similarly,

$$E(\bar{x}^2) = E(\bar{x} - \mu + \mu)^2 = E(\bar{x} - \mu)^2$$
$$+ E\{2\mu(\bar{x} - \mu)\} + E(\mu^2) \quad (5.27)$$

Since $E(\bar{x} - \mu) = E(\bar{x}) - \mu = \mu - \mu = 0$, and, see (5.23),

$$E(\bar{x} - \mu)^2 = \sigma^2/n,$$

then

$$E(\bar{x}^2) = \mu^2 + \sigma^2/n \tag{5.28}$$

Now $x_1, x_2, \ldots x_n$ are each independently selected at random, so by (5.26)

$$E(x_1^2) = E(x_2^2) = \ldots = E(x_n^2) = \sigma^2 + \mu^2 \tag{5.29}$$

Since the expected value of the sum of independent random variables is the sum of the separate expected values (formula (5.4)), then

$$E(x_1^2 + x_2^2 + \ldots + x_n^2 - n\bar{x}^2) = n(\sigma^2 + \mu^2)$$
$$- n(\sigma^2/n + \mu^2) = (n - 1)\sigma^2 \tag{5.30}$$

By formula (2.11), $s^2 = (x_1^2 + \ldots + x_n^2 - n\bar{x}^2)/(n - 1)$ hence

$$E(s^2) = \frac{(n - 1)\sigma^2}{n - 1} = \sigma^2 \tag{5.31}$$

The quantity at one time used to estimate σ^2 was

$$\{(x_1 - \bar{x})^2 + \ldots + (x_n - \bar{x})^2\}/n$$
$$= (x_1^2 + \ldots + x_n^2 - n\bar{x}^2)/n \tag{5.32}$$

a more natural quantity than s^2, but its expected value is $(n - 1)\sigma^2/n$, and so it is a 'biased' estimate of σ^2.

Any convenient quantity A can be subtracted from $x_1, x_2, \ldots x_n$ and provided it is subtracted from $\bar{x}$, we can use these 'coded' values to calculate s^2. That is

$$s^2 = \{(x - A)^2 + \ldots + (x_n - A)^2$$
$$- n(\bar{x} - A)^2\}/(n - 1) \tag{5.33}$$

There are other quantities which can be used to estimate σ^2, such as the range, but s^2 can be proved to be the most accurate.

Example

5.6(i). Calculate s^2 for each of the two samples: 0·19, 0·18, 0·21, 0·30, 0·66, 0·42, 0·08, 0·12, 0·30, 0·27; 0·15, 0·13, 0·00, 0·07, 0·27, 0·24, 0·19, 0·04, 0·08, 0·20, 0·12 (measurements of muscle glycogen in rabbits, the second set having been injected with insulin).

5.7 Distribution of sample means from a normal population, a non-normal population (the central limit theorem)

It is possible to prove mathematically that the sampling distribution of the mean $\bar{x}$, of samples of size n, is itself distributed normally.

The mean of the second normal distribution is also μ, and the variance is σ^2/n (as indeed must be the case since $E(\bar{x}) = \mu$ and var $(\bar{x}) = \sigma^2/n$ for any population, see Section 5.5). Its probability density function must therefore be

$$\frac{\sqrt{n}}{\sigma\sqrt{(2\pi)}} \exp\left[-\frac{n(\bar{x} - \mu)^2}{2\sigma^2}\right] \qquad (5.34)$$

Hence the probability that a sample mean lies between a and b is obtained by integrating (5.34) with respect to $\bar{x}$ from a to b.

The central limit theorem (which requires a rather elaborate proof) states that the population of sample means from a *non-normal* population approximates to the distribution (5.34), the approximation being better the larger n. This is a most useful result indeed and enables us to avoid the need in many problems, of classifying a population as normal, or otherwise.

5.8   The standard error of a statistic (may be omitted on a first reading)

A 'statistic' is any function of $x_1, x_2, \ldots x_n$; call this function $F(x_1, x_2, \ldots x_n)$. Now the expected, or mean, value of $F(x_1 \ldots x_n)$ is denoted by $E\{F(x_1, \ldots x_n)\}$. Then the 'variance' of $F(x_1, \ldots x_n)$ is, by definition,

$$\text{var}\,(F) = E\{F(x_1, \ldots x_n) - E[F(x_1, \ldots x_n)]\}^2 \qquad (5.35)$$

The square-root of var (F) is called the 'standard error' of $F(x_1, \ldots x_n)$. If F is used to estimate some quantity then the standard error gives some idea of the 'spread' of the estimates.

6

Applications of normal sampling theory; significance tests

6.1 Type of problem requiring such tests, meaning of 'significant'

A typical practical problem is this: It is desired to ascertain whether a dye affects the breaking strength of a type of cotton sewing thread. The breaking strength of a number of standard lengths of the dyed thread and a number of lengths of the undyed thread are measured. There will almost certainly be some difference between the mean strengths of the dyed and the undyed thread. But there is so much variability in cotton that the mean strengths of *two* samples of *undyed* thread will be different.

However, if the difference between the *dyed* and *undyed* mean strengths is *large* enough for such a difference to occur very *rarely indeed* between two samples of *undyed* thread, then we can be confident that there is very probably a real difference between dyed and undyed mean strengths (in which case we say there is a 'significant difference' between them).

In practically every scientific or technological measurement some not completely controlled factors are present (e.g. the chemical constituents are never 100% pure, readings of thermometers are not 100% accurate, etc.) and the consequent variability, though less than for cotton thread, may obscure change caused by the introduction of a new factor, etc. Hence the importance of tests which decide when differences are significant.

Such tests are called 'significance tests' and we now describe the most important of them, starting with the simpler cases. It is assumed that the populations involved are normal, but the following tests apart from that of Section 6.9 are fairly accurate for other distributions (because of the central limit theorem, see Section 5.7).

6.2 To test whether a single observation x could come from a normal population with given μ and σ (a two-tailed test)

Obviously the greater the difference between x and μ compared to the standard deviation σ the less likely is it that x is an individual from the given population. So we calculate the test-function

$$c = |x - \mu|/\sigma \qquad (6.1)$$

and if c is larger than a certain 'critical' value we decide that x is unlikely to come from the population.

The most commonly used 'critical value' for c is 1·96 because only 5% of individuals in a normal population have variate-values differing from μ by 1·96σ or more (see Fig. 6.1(a)). Thus, using the critical value for c of 1·96, we would make a mistake on 5% of occasions when x really did come from the population. The usual way of expressing this is to say that 'if $c > 1$·96 the difference between x and μ is significant at the 5% probability level'.

Consider this practical example. In a chemical plant it is known (from long experience) that the percentage of iron in the product is distributed normally with $\mu = 5$·37 and $\sigma = 0$·12. After a replacement to the plant the first subsequent test gives a reading $x = 5$·21. Here $c = |5$·$21 - 5$·$37|/0$·$12 = 1$·333 which is well below 1·96 and we would not think that anything was wrong even though there is a difference between x and μ.

Suppose, however, that this reading x was 5·69. Here $c = |5$·$69 - 5$·$37|/0$·$12 = 2$·667. This is greater than 2·58 which is another commonly used critical value of $c \equiv |x - \mu|/\sigma$, used because only 1% of individuals in a normal population have variate values differing from μ by 2·58σ or more (see Table 4.3). Hence, if an observed value of c is greater than 2·58 we say that 'the difference between x and μ is significant at the 1% probability level', meaning that we shall make a mistake on only 1% of occasions if x really did come from the population with mean μ and standard deviation σ. Hence a reading of $x = 5$·69 would cause us to get into touch immediately with the plant manager to tell him we thought that something was wrong.

Another commonly used critical value for c is 3·29 and if

$$c = |x - \mu|/\sigma > 3\text{·}29$$

we say that 'x and μ differ significantly at the 0·1% probability level' because we would make a mistake on 0·1% of occasions (i.e. 1 in a 1,000) if x really did come from the given population. Thus, suppose, immediately after the above-mentioned plant replacement our first reading x was 4·95. Here

$$c = |4\text{·}95\text{--}5\text{·}37|/0\text{·}12 = 0\text{·}42/0\text{·}12 = 3\text{·}50.$$

This is greater than 3·29 and we would be extremely confident that the plant was not functioning properly, though we would not be absolutely certain and there would be a slight probability (less than 1 in a 1,000) that we were wrong.

Beginners sometimes find this lack of absolute certainty a little upsetting. However they should note that nothing is certain, e.g. we go for a drive in the car though we know there is a chance of a fatal accident, similarly we cross the road though we are not *absolutely* sure of reaching the other side uninjured. We choose our probability level of accident according to the circumstances, thus, if we will make several thousand pounds we are prepared to travel quite fast from one engagement to another. Similarly, in applying the c-test, the critical value we use depends on the circumstances. In the chemical plant example quoted above we would weigh the loss of production caused by stopping the plant on the 5 % (or the 1 %, etc.) of occasions we were wrong, against the wasted production saved on the 95 % (or 99 % or, etc.) of occasions we were right.

Where the issue is not quite so easy to assess in terms of money the following will act as guide. Much experience has justified equating a probability level of 0·1 % with *very highly significant*, a level of 1 % with *highly significant* and a level of 5 % with *just significant* (note that the *lower* the probability level, i.e. the lower the probability of making a mistake, the *higher* the significance).

Example

6.2(i). Find whether the difference between x and μ is significant, and whether at the 5, 1, or 0·1 % probability level given that

$$x, \mu, \sigma = (a) \ 3\cdot21, \ 2\cdot64, \ 0\cdot32$$
$$(b) \ 6\cdot24, \ 69\cdot3, \ 3\cdot07$$
$$(c) \ 101\cdot4, \ 108\cdot1, \ 2\cdot6$$
$$(d) \ -2\cdot87, \ -1\cdot43, \ 0\cdot86.$$

6.2(ii). Test whether x is significantly different from μ for the data of Example 6.3(i) and (ii).

In the above test we decide that the difference between x and μ is significant either if x is greater than μ by a certain amount or if x is less than μ by the same amount. So it is called a 'two-tailed' test. It is the most commonly used form of test, but, occasionally, we are concerned with one-tailed tests (see Fig. 6.1 which illustrates the difference).

6.3 To test whether an observation x is significantly lower (or significantly greater) than μ (a one-tailed test)

There are occasions when we merely wish to know whether an observation x is significantly *lower* than μ (given, of course, the value of σ) and we do not mind if x is greater than μ, by however much. Thus, if we were dealing with the effect of dye on the strength of cotton yarn we might only be concerned if the strength was *lowered* (if it was increased we might in fact be pleased).

In this case we use the one-tailed test-function

$$c' = (\mu - x)/\sigma \qquad (6.2)$$

If c' is greater than the (one-tailed) critical value 1·645 we say that 'x is significantly *lower* than μ at the 5% probability level',

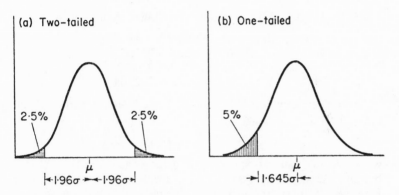

FIGURE 6.1. Two- and one-tailed tests at the 5% probability level

because we would be wrong on 5% of occasions if x really did come from a normal population with mean μ and standard deviation σ (see Fig. 6.1(a)). If c' is $<$1·645 or is *negative* we say that x is not significantly lower than μ at the 5% level. Table 6.1 gives critical values of c' at other probability levels (see also Table 4.3).

Table 6.1. CRITICAL VALUES OF c' OR c''

Critical value	1·645	1·96	2·33	2·58	3·09	3·29
Percentage probability level. .	5	2·5	1	0·5	0·1	0·05

If we wish to test whether x is significantly greater than μ then

$$c'' = (x - \mu)/\sigma$$

is our test function; it has the same critical values as c'.

Note that when a critical value, e.g. 1·96, is used for c' the probability level is *half* that for c (i.e. 2·5% for c' as against 5% for c).

42   ESSENTIALS OF STATISTICS

Examples

6.3(i). Test whether x is significantly greater than μ and at about what probability level if

$$x, \mu, \sigma = \quad (a)\ 8{\cdot}97,\ 7{\cdot}34,\ 0{\cdot}85$$
$$(b)\ 41{\cdot}3,\ 36{\cdot}4,\ 3{\cdot}5$$
$$(c)\ -7{\cdot}4,\ -9{\cdot}2,\ 0{\cdot}6$$
$$(d)\ 2{\cdot}1,\ -1{\cdot}8,\ 1{\cdot}3.$$

6.3(ii). Test whether x is significantly lower than μ when

$$x, \mu, \sigma = \quad (a)\ 3{\cdot}04,\ 5{\cdot}97,\ 1{\cdot}64$$
$$(b)\ -5{\cdot}23,\ -3{\cdot}17,\ 0{\cdot}45$$
$$(c)\ 11{\cdot}87,\ 14{\cdot}13,\ 3{\cdot}92$$
$$(d)\ -8{\cdot}17,\ -5{\cdot}32,\ 0{\cdot}67.$$

The reader may feel that this one-tailed test in some manner contradicts the two-tailed test of Section 6.2 in that if

$$(x - \mu)/\sigma = 1{\cdot}645$$

we would say that x is significantly *greater* than μ at the 5% level but since $|x - \mu|/\sigma < 1{\cdot}96$, x and μ are not significantly *different*. However, it must be noted that the tests are used for *different* circumstances and that the reader must decide before the observations are made if he is concerned with whether x is merely greater than, or whether x is different from μ. In any case in practical applications the loss caused on the 5 or 1% of occasions when the test gives a wrong result decides which test is to be used.

6.4 To test whether a sample mean $\bar{x}$ differs significantly from μ (with σ known)

In practice we do not rely on one observation if several (i.e. a sample) can be obtained without trouble. So we now consider the test for a sample. Now we know (see Section 5.7) that, for a sample of size n from a normal population, the mean $\bar{x}$ is itself distributed normally with (the same) mean μ but with standard deviation $\sigma/\sqrt{n}$ (this is also true to a fairly good approximation even if the population is *not* normal, see Section 5.7).

Our (two-tailed) test function is, therefore,

$$c = |\bar{x} - \mu|/(\sigma/\sqrt{n}) \tag{6.3}$$

which has the same critical values as the c of 6.2. (If we were only concerned with whether $\bar{x}$ was significantly lower than μ we would use the one-tailed test-function

$$c' = (\mu - \bar{x})/(\sigma/\sqrt{n})$$

which has the critical values of Table 6.1.) Thus, if say, $\bar{x} = 10\cdot3$, $\mu = 14\cdot2$, $\sigma = 4\cdot8$, $n = 9$ then

$$c = 3\cdot9/(4\cdot8/3) = 3\cdot9/1\cdot6 = 2\cdot44$$

which is greater than $1\cdot96$ and so the difference between $\bar{x}$ and μ is significant at the 5% level (though not at the 1% level).

Examples

6.4(i). Find whether $\bar{x}$ and μ differ significantly and state at what probability level, when

$$\mu, \sigma, \bar{x}, n = \begin{array}{l} (a) \ 4\cdot39, \ 0\cdot42, \ 4\cdot56, \ 25 \\ (b) \ -1\cdot39, \ 1\cdot62, \ -0\cdot45, \ 16 \\ (c) \ 0\cdot046, \ 0\cdot011, \ 0\cdot035, \ 8 \\ (d) \ 79\cdot6, \ 5\cdot43, \ 76\cdot4, \ 6. \end{array}$$

6.4(ii). Find whether $\bar{x}$ and μ differ significantly and state at what probability level, when $\mu = 5\cdot37$, $\sigma = 0\cdot12$ and the observations are $5\cdot19$, $5\cdot17$, $5\cdot24$, 5.12.

6.4(iii). Find whether $\bar{x}$ and μ differ significantly and state at what probability level, when $\mu = 60\cdot4$, $\sigma = 2\cdot1$ and the observations are $60\cdot9$, $63\cdot8$, $61\cdot2$, $6\cdot32$, $60\cdot5$, $60\cdot8$.

6.4(iv) Test if $\bar{x}$ is significantly greater than μ for the data of Examples 6.4(i) (a) and (b), 6.4(ii), and 6.4(iii).

6.4(v). Test if $\bar{x}$ is significantly lower than μ for the data of Examples 6.4(i) (b), (c), (d), 6.4(ii).

6.4(vi). A manufacturer claims that his thread has a mean breaking strength of 142 grammes with a standard deviation of $8\cdot6$ grammes. Does the mean of these observed breaking strengths differ significantly from the claimed value: 126, 150, 133, 118, 124, 139, 131, 141, 128, 136, 130, 137?

If the observed value of c, though below, is not far from the critical value of c (at the desired probability level), then more observations should be taken (if possible). The relation between what the reader knows or thinks is a real physical difference and the critical statistical difference (i.e. $1\cdot96\sigma/\sqrt{n}$, at the 5% level) should be noted. If the latter is much larger than the former then a sample of much larger size should be taken. Note, too, that the critical difference $1\cdot96\sigma/\sqrt{n}$ falls off inversely as the square root of n. So that doubling n only lowers it to $1/\sqrt{2} = 0\cdot707$ of its previous value, and taking n a hundred times larger only reduces this critical difference to $1/10$ of its former value. If, however, the critical statistical difference is well below the physical difference there is no point in increasing n.

6.5   To test the difference between two sample means with σ known

This is not a common case but it illustrates certain points. (Note that we do not know μ here and in fact we are not directly interested in μ, only in whether the difference between the sample means is significant when compared to σ.) Suppose $\bar{x}_1$ is the mean of a sample of size n_1, and $\bar{x}_2$ is the mean of a sample of size n_2. We know that the distribution of $\bar{x}_1$ is normal with variance σ^2/n_1, and that of $\bar{x}_2$ is normal with variance σ^2/n_2 (see Section 5.7). If they are samples from the *same* population then, by the results of Section 4.5, the difference $\bar{x}_1 - \bar{x}_2$ is distributed normally with *mean* $\mu - \mu = 0$, and *variance* $\sigma^2/n_1 + \sigma^2/n_2$. Hence our test-function is

$$
\begin{aligned}
c &= \frac{|\bar{x}_1 - \bar{x}_2| - 0}{\sqrt{(\sigma^2/n_1 + \sigma^2/n_2)}} \\
&= \frac{|\bar{x}_1 - \bar{x}_2|}{\sigma\sqrt{(1/n_1 + 1/n_2)}}
\end{aligned}
\tag{6.4}
$$

As before, if $c > 1.96$ or 2.58 or 3.29, we say the difference $\bar{x}_1 - \bar{x}_2$ is significant at the 5 or 1 or 0.1% probability level.

Examples

6.5(i). Given $\bar{x}_1 = 14.97$, $\bar{x}_2 = 15.63$, $n_1 = 10$, $n_2 = 20$, $\sigma = 0.65$ test for significance.

6.5(ii). Assume that for European races men's heights have a standard deviation $\sigma = 2.51$ in., and that the mean height of 6,914 Englishmen is 67.44 in., and of 1,304 Scotsmen is 68.55 in.; test the difference.

6.6   To test whether a sample mean $\bar{x}$ differs significantly from a given population mean μ when σ^2 is estimated from the sample

In most practical problems the value of σ^2 is *not known* beforehand and (see Section 2.5) is usually estimated by the formula

$$
s^2 = \{(x_1 - \bar{x})^2 + (x_2 - \bar{x})^2 + \ldots + (x_n - \bar{x})^2\}/(n - 1) \tag{6.5}
$$

Replacing σ by s in (6.1) we have the modified test-function

$$
t = |\bar{x} - \mu|/(s/\sqrt{n}) \tag{6.6}
$$

At one time the same critical values (1.96, etc.) were used for t as for c; W. S. Gosset, who wrote under the name 'Student', first realized this was incorrect; later, Fisher worked out the sampling distribution of the 'statistic' $(\bar{x} - \mu)/(s/\sqrt{n})$ and this enabled the correct critical values to be calculated. Some of these are given in Table 6.2 and a fuller table in the Appendix (Table A3). Note that these critical values depend on a quantity ν called the 'degrees of

Table 6.2. SOME CRITICAL VALUES OF t

Degrees of freedom v			1	2	4	6	9	20	∞
	One-tailed	Two-tailed							
Percentage .	2·5	5	12·71	4·30	2·78	2·45	2·26	2·09	1·06
Probability .	0·5	1	63·66	9·93	4·60	3·71	3·25	2·85	2·58
Level .	0·05	0·1	636·60	31·60	8·61	5·96	4·78	3·85	3·29

freedom' which is, in general, the same as the denominator in the formula for s^2. We see from formula (6.5) that this denominator, and, hence, the degrees of freedom $v = n - 1$ (in other applications the denominator and, hence, v is different, see Section 6.8). It will be noted that as v approaches ∞, the critical values of t approach those of c.

As with the c test of the previous sections, if we take the 5 or 1 or 0·1 % probability level values of t as 'critical value', we shall make a mistake on 5 or 1 or 0·1 % of occasions when the sample really does come from a population with the given value of μ.

Suppose for example, that $\bar{x} = 15\cdot7$, $\mu = 18\cdot3$, $s = 4\cdot0$, $n = 10$. Here, $t = 2\cdot6/(4/\sqrt{10}) = 2\cdot05$. Now $v = n - 1 = 9$ and the (two-tailed) critical value for t for $Y = 9$ at the 5 % level is 2·26. So the difference is *not* significant at this level.

Again, if we are only concerned with whether $\bar{x}$ is greater than (or smaller than) μ we use the one-tailed test function $t = (\bar{x} - \mu)/(s/\sqrt{n})$, (or $t = (\mu - \bar{x})/(s/\sqrt{n})$). The one-tailed probability levels are half those of the two-tailed.

Examples

6.6(i). Is the difference between $\bar{x}$ and μ significant and at about what probability level when

$$\bar{x}, \mu, s, n = (a) \ 6\cdot84, 7\cdot15, 0\cdot36, 10$$
$$(b) \ -4\cdot32, -5\cdot17, 0\cdot37, 7$$
$$(c) \ 101\cdot4, 99\cdot7, 2\cdot8, 21.$$

6.6(ii). For a certain chemical product theory gives 3·44 as the correct percentage of calcium. Ten analyses gave 3·05, 2·97, 2·94, 3·29, 2·98, 3·05, 2·88, 3·51, 3·03, 3·25; is their mean significantly different from 3·44?

6.6(iii). Test whether $\bar{x}$ is significantly greater than μ for the data of 6.6(i) (*a*), (*b*), (*c*).

The t-test applied to 'paired comparisons'

The test of Section 6.5 is often used in a way illustrated by the following: In an experiment on the effect of electric current on the height of maize seedlings, ten pairs of boxes of seedlings were put in various places and a weak electric current applied to one box of each pair (pairing the boxes avoids errors likely to arise from putting the electrified boxes in different places from the unelectrified boxes, in which case any observed difference might be due to the different environments). The average height of seedlings in each box was measured and the average electrified — average unelectrified height in mm for each box pair was: 6·0, 1·3, 10·2, 23·9, 3·1, 6·8, −1·5, −14·7, −3·3, 11·1. The mean $\bar{x}$ is 4·29, and s^2 is 104·1. We wish to test the hypothesis that the electric current has no effect, i.e. the hypothesis that $\mu = 0$. Here

$$t = |\bar{x} - \mu|/(s/\sqrt{n}) = (4·29 - 0)/(\sqrt{104·1}/\sqrt{10}) = 1·33.$$

This is appreciably less than the critical value of t at the 5% level, which, with $\nu = n - 1 = 9$ is 2·26 (see Table 6.2). Thus there is no reason to reject the hypothesis that $\mu = 0$, and the results were considered to indicate that the electric current had no effect.

Note that in 'paired comparison' experiments the *difference* between each pair of readings constitutes the observation when the t test is applied to test the hypothesis that $\mu = 0$ (this hypothesis is often called the 'null hypothesis').

Examples

6.7(i). Test the mean of 3, 6, −9, −13, −4, 1, −8, −7, 2; for significant difference from $\mu = 0$.

6.7(ii). A paired comparisons experiment gave the pairs of readings 62·5, 51·7; 65·2, 54·2; 67·6, 53·3; 69·9, 57·0; 69·4, 56·4; test the null hypothesis.

6.7(iii). Counts by Peterson in 1954 of the number of seeds per pod of top and bottom flowers of lucerne were (top first) 4·0, 4·4; 5·2, 3·7; 5·7, 4·7; 4·2, 2·8; 4·8, 4·2; 3·9, 4·3; 4·1, 3·5; 3·0, 3·7; 4·6, 3·1; 6·8, 1·9. Test the null hypothesis (i.e. the hypothesis that there is no significant difference between top and bottom counts).

6.8 To test the difference between the means of any two samples

This is the most used test of this chapter because we rarely know the exact variances of the two populations from which the samples are taken. Further, as will be shown in Section 6.8(*c*), by the Welch-test we need know *nothing* about them (other than what information is contained in the observations in each sample).

(a) **Notation.** Suppose n, n' are the sizes of the two samples and that x_1, x_2, . . . x_n, and x_1', x_2', . . . $x_{n'}'$, are the two sets of observations. Let $\bar{x}$, $\bar{x}'$ and s^2, $(s')^2$ denote the sample means and variances; that is

$$\bar{x} = (x_1 + x_2 + \ldots + x_n)/n, \tag{6.7}$$
$$\bar{x}' = (x_1' + x_2' + \ldots + x_{n'}')/n'$$

$$s^2 = \frac{(x_1 - \bar{x})^2 + \ldots + (x_n - \bar{x})^2}{n - 1},$$
$$s'^2 = \frac{(x_1' - \bar{x}')^2 + \ldots + (x_{n'}' - \bar{x}')^2}{n' - 1} \tag{6.8}$$

The quantity S is used in the test of Section 6.8(b) where

$$S^2 = \{[(x_1 - \bar{x})^2 + \ldots + (x_n - \bar{x})^2]$$
$$+ [(x_1' - \bar{x}')^2 + \ldots + (x_{n'}' - \bar{x}')^2]\}/(n + n' - 2) \tag{6.9}$$

N.B. $S^2 = [(n - 1)s^2 + (n' - 1)(s')^2]/(n - 1 + n' - 1) \tag{6.10}$

(b) **Test, when population variances are equal (t-Test)**

If we know that the samples come from two populations whose (true) variances are equal (though the population means will be different) then it can be proved that this common variance σ^2 is best estimated by S^2. Replacing σ by S in (6.4), we see that the appropriate (two-tailed) test function is

$$t = \frac{|\bar{x} - \bar{x}'|}{\sqrt{\{S^2(1/n + 1/n')\}}} \tag{6.11}$$

which has the critical values of Table 6.2 and Table A.3, but whose 'degrees of freedom' $\nu = n + n' - 2$ (i.e. the denominator of S^2, see 6.9).

Suppose, for example, that $\bar{x}$, $\bar{x}'$, n, n', $S^2 = 107, 123, 8, 14, 154$ then $t = 16/\sqrt{\{154 \times (22/112)\}} = 64/22 = 2\cdot919$. Now critical t for $\nu = 8 + 14 - 2 = 20$ at the 1% level is $2\cdot84$, so the difference between $\bar{x}$ and $\bar{x}'$ is significant at this level, since observed t is greater than critical t. (If we were using the *one-tailed* test, $\bar{x}'$ would be significantly *greater* than $\bar{x}$ at the $0\cdot5\%$ level, see Table A.3.)

Examples

6.8(i). Test for significant difference the following:

$$\bar{x}, \bar{x}', n, n', S^2 = (a)\ 47\cdot3, 56\cdot6, 16, 9, 97\cdot4$$
$$(b)\ -0\cdot173, -0\cdot114, 100, 100, 0\cdot0263$$
$$\bar{x}, \bar{x}', n, n', s^2, (s')^2 = (c)\ 102, 110, 9, 11, 86, 96$$
$$(d)\ 16\cdot5, 14\cdot8, 10, 20, 46, 103.$$

6.8(ii). Test the difference between the means of 15, 18, 17, 21, 19, 24, 18; 10, 8, 16, 11, 13, 19, 14, 13, 11, 10.

6.8(iii). Test whether $\bar{x}'$ is significantly greater than $\bar{x}$ in Example 6.8(i) (b), (c), and (d).

Unfortunately we rarely know that the two population variances are the same and the test function (6.11), which was once commonly used, is being superseded by the following perfectly general test.

(c) The Welch-Test, when population variances may differ.

Welch has given a remarkable solution to this problem and his test applies to all cases whether the variances are the *same* or whether they *differ*, no matter by *how much*. The test is, in fact, independent of *any* previous knowledge of the population variances and is thus universal. The procedure is as follows: The (two-tailed) test-function W is

$$W \equiv |\bar{x} - \bar{x}'|/\sqrt{\{s^2/n + (s')^2/n'\}} \tag{6.12}$$

The quantity

$$(s^2/n)/\{s^2/n + (s')^2/n'\} \equiv h$$

is also calculated. Critical values of W for various values of $n - 1$, $n' - 1$, and h are given in the Appendix, Table A.4. Note that these critical values depend on h, as well as $n - 1$ and $n' - 1$ (indeed it was by making them depend on h that Welch obtained this very general solution).

For example suppose that $\bar{x}$, $\bar{x}'$, n, n', s^2, $(s')^2 = 453$, 475, 9, 11, 540, 440. Then $W = 22/\sqrt{(540/9 + 440/11)} = 2.2$ and

$$h = 60/(60 + 40) = 0.6.$$

Table A.4 gives critical W at 5% probability level for $n - 1 = 8$, $n' - 1 = 10$, $h = 0.6$ to be 2.08, so the difference between x and x' is significant at the 5% (since observed $W >$ critical W) level.

For the one-tailed test to tell whether $\bar{x}$ is significantly greater than $\bar{x}'$ we use $W' = (\bar{x} - \bar{x}')/\sqrt{\{s^2/n + (s')^2/n'\}}$; the critical values of W' are also given in Table A.4.

Example

6.8(iv). Test, by the Welch test, the data of Examples 6.8(i) (c), (d) and 6.8(ii), for (a) significant difference between means, (b) whether the first mean is significantly greater than the second.

6.9 To test whether the observed variances of two samples differ significantly (by the F test)

Here we test not the *difference* between the two observed variances but their *ratio*. That is, if s_1^2 and s_2^2 are the two sample variances (calculated by the s^2 formula (2.10)), our test-function is

$$F = s_1^2/s_2^2 \qquad (6.13)$$

where s_1^2 must always be the *larger* variance (hence F is always larger than unity). Table A.6 of the Appendix gives critical values for F, note that these critical values depend: (a) on the probability level, (b) on the degrees of freedom v_1 with which s_1^2 is calculated and so, if n_1 is the size of the sample, $v_1 = n_1 - 1$, and (c) on the degrees of freedom v_2 of the smaller variance s_2^2, i.e. $v_2 = n_2 - 1$ where n_2 is the size of the sample. (Chapter 7 gives uses of the F test where v_1 and v_2 have different values.)

Consider the following case: $s_1^2 = 8 \cdot 73$, $n_1 = 11$, $s_2^2 = 1 \cdot 89$, $n_2 = 8$. Here, observed $F = 8 \cdot 73/1 \cdot 89 = 4 \cdot 62$; $v_1 = 11 - 1 = 10$, $v_2 = 8 - 1 = 7$; and critical F for $v_1 = 10$, $v_2 = 7$ is $3 \cdot 64$ at the 5% level, and is $6 \cdot 62$ at the 1% level (see Table A.6). So there is a significant difference at the 5%, but not at the 1% level, since observed $F >$ critical F at the former though not the latter level.

The critical values of F are based on the calculations of the distribution of the ratio of two sample variances from a normal population made by Sir R. A. Fisher—hence the name F test.

Examples

6.9(i). Test by the F test the following

$$s_1^2, s_2^2, n_1, n_2 = (a)\ \ 0 \cdot 793,\ 0 \cdot 137,\ 7,\ 9$$
$$(b)\ \ 703,\ 85,\ 3,\ 8$$
$$(c)\ \ 11 \cdot 3,\ 4 \cdot 8,\ 10 \cdot 6.$$

6.9(ii). Test the ratio of the variances of the samples given in Examples 6.7(ii), in 6.7(iii), and in 6.8(ii).

By taking s_1^2 to be the larger variance we have made what is really a two-tailed test (which would indicate significance when s_1^2 was either much smaller than, or much larger than s_2^2) into an apparently one-tailed test. This need not bother the reader in practice, however.

6.10 Some notes on significance tests

(a) **Interpretation of non-significance.** If a test gives a non-significant answer, it must be remembered that this is *statistical non-significance* and that there may be present some physical

50   ESSENTIALS OF STATISTICS

difference which is too small relative to the standard deviation to be revealed (unless perhaps more observations are made). It is best to regard a *non-significant* result of finding a statistical difference (especially if near the border-line value) as equivalent to the Scottish verdict *non-proven*.

If it is decided to take more observations and these are added to the original, then the above tests should be used with a little more caution. Such a procedure is known as a 'sequential' procedure, and some tables by Arnold and Goldberg, National Bureau of Standards (1951), Washington, show how to do this accurately.

(*b*) **Two types of error.** As mentioned previously there is always a probability, albeit small, that though the test indicates a significant difference the two populations from which the samples come really have the same mean (or variance in the case of the *F* test of Section 6.9). Such an error is called a Type I error.

However, there is another possible error, the Type II error, which arises when the test indicates *no* significant difference though the population means *really* are different.

In most practical applications we usually wish to be sure that when there is a real difference it is detected and so we are concerned with making the Type I error as small as we reasonably can.

(*c*) **Interpolation when using tables.** It is not practicable to give critical values of say, *t* for every value from 1 to ∞ of *ν* the degrees of freedom. The tables provide, however, sufficient information for critical *t* to be found for those values of *ν* not given explicitly. This we usually do in statistics by taking a 'proportional' value; an illustration will make this clear. Now critical *t* in Table 6.2 at the 5% level for *ν* = 6 is 2·45 and for *ν* = 9 is 2·26. The proportional value for *ν* = 8 would be 2·26 + (2·45 − 2·26) × ⅓ = 2·32 correct to 2 decimal places. (Table A.3 gives 2·31; however, such slight differences do not seriously affect practical applications.) For large values of *ν* we have to take proportional parts with respect to 1/*ν*. Thus to find critical *t* for *ν* = 50 at the 1% level from Table 6.2 we note that *t* = 2·85 for *ν* = 20 when 1/*ν* = 0·05, and 2·58 for *ν* = ∞ when 1/∞ = 0. Hence, for *ν* = 50 when 1/*ν* = 0·02,

$$t = 2·58 + (2·85 − 2·58) × 0·02/0·05 = 2·69$$

(in fact 2·68 is the correct answer). It is possible to use more elaborate methods than this taking of proportional values (which is known, technically as 'linear interpolation') and thus to get greater accuracy, but it is not worth the extra effort in practical statistical applications.

For critical *W* values (Table A.4) we may have to 'interpolate' with respect to *n* − 1, *n'* − 1, and *h* (or if *n* − 1 or *n'* − 1 is large with respect to 1/(*n* − 1) or 1/(*n'* − 1)). However, since difference between successive W values in the table are all small this triple interpolation will not lead to any practical loss of accuracy. In other tables, e.g. the F table we may have to doubly interpolate (with respect to both v_1 and v_2). However, the simple proportional part technique will suffice for practical purposes.

(*d*) **Testing a hypothesis.** This is merely an alternative way of stating a significance test, which has certain advantages. Thus we can restate the aim of 6.2 as 'to test the hypothesis that the observation x could come from a normal population with given μ and σ'. If $c = |x - \mu|/\sigma$ is larger than the critical value at, say, the 5% level we say 'we reject the hypothesis at the 5% level'. If c is less than this critical value we say 'we accept the hypothesis'; this means that we think μ, σ are possible values for the population mean and standard deviation not that they are *the only* possible values.

7

Normal sampling theory: test for difference between several sample means, analysis of variance, design of experiments

7.1 Nature of the test for difference between several sample means

In Chapter 6 the t test was used to decide whether the difference between two sample means was too great for them to come from the same population. We now describe the test for the case of more than two samples (Note: applying the t test to the difference between largest and smallest mean will give the wrong answer). This test is based on the fact that the mean of a sample of size n from a normal population with given σ, is itself normally distributed with variance σ^2/n (see Section 5.7). If, then, we have m samples, and their means have a greater variance than would be expected from m individuals from a population whose variance is σ^2/n, we conclude that, very probably, there are some real physical differences between the means. Unfortunately, σ^2 is not usually known and so has to be estimated from the variance within each sample (it is assumed that *all* population variances are the same).

Hence, for example, if we wish to test the effect of several different dyes on the breaking strength of cotton thread, we take a sample of n observations of breaking strength from each kind of dyed thread, including the undyed thread. Suppose there are m samples in all. If the dyes have no effect then the observed variance of their means $(s')^2$, given by formula 7.1, is an estimate of σ^2/n with $m-1$ degrees of freedom. But the observed variance within a single sample is also an estimate of σ^2 with $n-1$ degrees of freedom. We can average these estimates of σ^2 from the m samples to get an overall estimate $(s'')^2$, given by formula 7.2, with $m(n-1)$ degrees of freedom. We then compare the two estimates $n(s')^2$ and $(s'')^2$ for σ^2, by the F test. If $n(s')^2$ is significantly larger than $(s'')^2$, then real differences are probably present.

The set of observations are usually tabulated as in Table 7.1. Note that: (i) we use the more general term 'treatment' instead of dye, (ii) each dye is referred to as a different 'level' of treatment, (iii) the jth observation at the ith level of treatment is denoted by x_{ij}, (iv) the levels may be numbered in any order. Note also that each 'treatment level' may differ qualitatively from the others, e.g.

they may be blue, green, red, etc. dyes of different chemical composition, and they are not necessarily different quantities of the same dye.

Table 7.1. OBSERVATIONS ON m SAMPLES (CLASSIFIED ONE-WAY)

Treatment level	: 1	2	. . . m	
	1	x_{11}	$x_{21} . . . x_{m_1}$	Grand mean
Observation	. 2	x_{12}	$x_{22} . . . x_{m_2}$	$\bar{x} =$
Number:	. .	.	. . .	$(\bar{x}_1 + \bar{x}_2 + . . . + \bar{x}_m)/m$
	n	x_{1n}	$x_{2n} . . . x_{mn}$	
Mean	: $\bar{x}_1$	$\bar{x}_2 . . . \bar{x}_m$		

Note: the x_{ij} may be 'coded' by subtracting A from each of them.

Suppose, for example, we have four differently dyed specimens of the same cotton thread, and that the following is the result of taking a sample of 3 breaking strength measurements on each specimen. The means of each sample and the grand mean will then be as shown.

Dye Colour (= Treatment Level:	Red 1	Green 2	Blue 3	Undyed 4)
Observation 1	104	104	105	114
Number: 2	102	106	103	111
3	100	105	104	108

Sample Means: $\bar{x}_1 = 102$ $\bar{x}_2 = 105$ $\bar{x}_3 = 104$ $\bar{x}_4 = 111$
Grand Mean: $\bar{x} = (102 + 105 + 104 + 111)/4 = 105.5$

7.2 Test for difference between means of samples of the same size

Given the set of observations of Table 7.1, the mean of each column is calculated, then the 'grand' mean $\bar{x}$ of these means, and, finally, $(s')^2$ by (7.1) and $(s'')^2$ by (7.2).

$$(s')^2 = \{(\bar{x}_1 - \bar{x})^2 + (\bar{x}_2 - \bar{x})^2 + . . . + (\bar{x}_m - \bar{x})^2\}/(m - 1)$$

$$\equiv \left\{\sum_{i=1}^{m} (\bar{x}_1 - \bar{x})^2\right\}/(m - 1) \qquad (7.1)$$

$$(s'')^2 = \{[(x_{11} - \bar{x}_1)^2 + (x_{12} - \bar{x}_1)^2 + . . . + (x_{1n} - \bar{x})^2]$$
$$+ [(x_{21} - \bar{x}_2)^2 + . . . + (x_{2n} - \bar{x}_2)^2] + . . .$$
$$+ [(x_{m1} - \bar{x}_m)^2 + . . . + (x_{mn} - \bar{x}_m)^2]\}/[m(n-1)]$$

$$\equiv \left[\sum_{i=1}^{m}\sum_{j=1}^{n} (x_{ij} - \bar{x}_i)^2\right]/[m(n - 1)] \qquad (7.2)$$

Since $(s')^2$ is an estimate of σ^2/n with $m - 1$ degrees of freedom, i.e. $n(s')^2$ estimates σ^2 with $m - 1$ degrees of freedom, and $(s'')^2$ is an estimate of σ^2 with $m(n - 1)$ degrees of freedom, we calculate the test-function (see Section 6.9)

$$F = n(s')^2/(s'')^2 \tag{7.3}$$

If this observed F is greater than critical F with $v_1 = m - 1$ and $v_2 = m(n - 1)$, at the desired probability level (Table A6, Appendix) we conclude that some, at least, of the treatment levels are causing significant variation. (In the case of the dyed threads of 7.1 if the observed F is significantly large, we can then test each individual dyed thread against the undyed thread by the t test, to see which are affecting breaking strength; but if the observed F is not significantly large then we can be sure that the t test will not show significance between any pair.)

Consider the breaking strength data at the end of Section 7.1. Here $m = 4$, $n = 3$, and, applying (7.1), (7.2), and (7.3) we have

$$(s')^2 = \{(-3 \cdot 5)^2 + (-0 \cdot 5)^2 + (-1 \cdot 5)^2 + (5 \cdot 5)^2\}/4 = 11 \cdot 25$$

$$(s'')^2 = \{(104 - 102)^2 + (102 - 102)^2 + (100 - 102)^2$$
$$+ (104 - 105)^2 + \cdots + (114 - 111)^2\}/8 = 3 \cdot 75$$

Hence $F = 3 \times 11 \cdot 25/3 \cdot 75 = 9 \cdot 0 > \text{crit } F(v_1 = 3, \ v_2 = 8) = 7 \cdot 59$ at the 1% level (see Table A6). So there is significant difference between sample means here. The test of 6.8(b) shows that each dyed mean differs significantly from the undyed mean at the 5% level. However, the dyed means do not differ significantly among themselves.

The reader must note that the term 'treatment' is very general indeed; for example, if the observations are counts of the number of tomatoes on each of n plants (selected at random) from each of m varieties of tomato plant, then, each variety is a different 'treatment level'. Or, if the observations are analyses of the product of a factory, n analyses being made each day of the week, to see if there is variation from day to day, then the days of the week are the treatment levels.

Note also that the observations in a single column of Table 7.1 may be rearranged in any order, but that the observations in a single row *cannot* be rearranged without altering the means. Hence the observations are said to be classified in 'one-way' only.

Example

7.2(i). Test the differences between the means of the following 5 samples: 8, 7, 9; 10, 8, 6; 7, 4, 4; 13, 16, 13; 9, 4, 2.

7.3 'Analysis of variance' for (one-way classified) samples of equal size

Sometimes the 'treatment' is a substance used in making a product, where the substance is subject to variation (no commercial substance is 100% pure). The m treatment levels are now m different specimens of the substance (e.g. from m different manufacturers). The aim of the investigation is first to see whether the different specimens produce significant variation in the final product (by the above test of Section 7.2); and, if so, to estimate, in addition, what part of the variance of the final product is due to this substance (i.e. we 'analyse the variance' of the product into parts, hence the name 'analysis of variance'). So we make the product using a specimen of the substance and take n observations (of whatever property is of importance, e.g. viscosity if the product is oil); then repeat for each specimen in turn.

The observations are set out as in Table 7.1 (each specimen is a different treatment level) and the calculations of Section 7.2 are made. However, this time $(s')^2$ and $(s'')^2$ have slightly different properties. If σ_b^2 is the variance in the product which is due to the substance and if σ^2 is the variance due to all other causes, then $(s')^2$, formula (7.1), is an estimate of $\sigma_b^2 + \sigma^2/n$ (see Section 4.5); while $(s'')^2$ is an estimate of σ^2 only.

We then test the hypothesis that $\sigma_b = 0$ by, as in Section 7.2, testing $n(s')^2/(s'')^2 \equiv$ observed F, against critical F with $\nu_1 = m - 1$, $\nu_2 = m(n - 1)$, using Table A.6. If observed F is greater than critical F (at the desired probability level) we conclude that σ_b^2 cannot be zero, and we estimate it from the formula

$$\sigma_b^2 = (s')^2 - (s'')^2/n \qquad (7.4)$$

If the observed F is not significantly large then σ_b^2 can be neglected and the theoretical best estimate of σ^2 is $\sum\sum(x_{ij} - \bar{x})^2/(mn - 1)$, though this usually differs little from $(s'')^2$ in practice.

This 'analysis of variance' is conveniently set out in Table 7.2.

Note the following formulae which are useful as checks and/or which simplify the labour of calculation if desk machines are available:

$$(m - 1)(s')^2 = \sum(\bar{x}_i - \bar{x})^2 = \bar{x}_1^2 + \bar{x}_2^2 + \ldots + \bar{x}_m^2 - m\bar{x}^2 \quad (7.5)$$

$$m(n - 1)(s'')^2 = \sum\sum(x_{ij} - \bar{x})^2; \qquad (7.6)$$

$$\sum\sum x_{ij}^2 - mn\bar{x}^2 = \sum\sum(x_{ij} - \bar{x})^2;$$

$$\sum\sum(x_{ij} - \bar{x})^2 = n\sum(\bar{x}_i - \bar{x})^2 + \sum\sum(x_{ij} - \bar{x}_i)^2; \quad (7.7)$$

$$m - 1 + m(n - 1) = mn - 1$$

Table 7.2. ANALYSIS OF VARIANCE FOR ONE-WAY CLASSIFICATION

Source of variation	Sum of squares S	Degrees of Freedom D	S/D estimates
Between samples	$n \sum\limits_{i=1}^{m} (\bar{x}_i - \bar{x})^2$	$m - 1$	$n\sigma_b^2 + \sigma^2$
Within samples	$\sum\limits_{i=1}^{m} \sum\limits_{j=1}^{n} (x_{ij} - \bar{x}_i)^2$	$m(n - 1)$	σ^2
Total	$\sum\limits_{i=1}^{m} \sum\limits_{j=1}^{n} (x_{ij} - \bar{x})^2$	$mn - 1$	

In words, (7.7) means that the sum of squares between samples and within samples add to give the total sum of squares; the same is true of the degrees of freedom column.

Examples

7.3(i). Establish the significance or otherwise of σ_b^2 and calculate estimates of σ^2, σ_b^2 when

$$n\sum(\bar{x}_i - \bar{x})^2 = 25 \cdot 7, \quad \sum\sum(x_{ij} - \bar{x})^2 = 74 \cdot 3;$$
$$m = 3, \quad n = 4.$$

(N.B. $\sum(x_{ij} - \bar{x})^2 = 74 \cdot 3 - 25 \cdot 7 = 48 \cdot 6$.)

7.3(ii). Establish the significance or otherwise of σ_b^2 and calculate estimates of σ^2, σ_b^2 when

$$n\sum(\bar{x}_i - \bar{x})^2 = 24 \cdot 4, \quad \sum\sum(x_{ij} - \bar{x})^2 = 51 \cdot 6; \quad m = 9, \quad n = 5$$

7.3(iii). Establish the significance or otherwise of σ_b^2 and calculate estimates of σ^2, σ_b^2 when the observations are

11	6	13	18
8	6	12	14
8	3	8	13

7.3(iv). Establish the significance or otherwise of σ_b^2 and calculate estimates of σ^2, σ_b^2 when the observations are

9	14	8	16	13
10	9	6	15	10
5	7	4	14	10

We usually call σ^2 the 'residual' variance because it is due to all sources of variation other than that due to the treatment (which is σ_b^2).

7.4 Analysis of variance for (one-way classified) samples of unequal size

Suppose that (i) the m samples are of size $n_1, n_2 \ldots n_m$, (ii) the sums of the observations in each sample are $S_1, S_2, \ldots S_m$, (iii) the total number of observations is N and the total sum of the observations is T, i.e.

$$n_1 + n_2 + \ldots + n_m = N, \quad S_1 + S_2 + \ldots + S_m = T \quad (7.8)$$

We then calculate

$$V_b = \frac{\sum\limits_{i=1}^{m} S_i^2/n_1 - T^2/n}{m - 1}, \quad V_r = \frac{\sum\limits_{i=1}^{m}\sum\limits_{j=1}^{n_i} x_{ij}^2 - \sum\limits_{i=1}^{m} S_i^2/n_i}{N - m} \quad (7.9)$$

where V_b and V_r are estimates of the variances shown

$$V_b \to \sigma^2 + \frac{[N^2 - (n_1^2 + n_2^2 + \ldots + n_m^2)]\sigma_b^2}{N(m - 1)}, \quad V_r \to \sigma^2 \quad (7.10)$$

V_b having $m - 1$, and V_r having $N - m$ degrees of freedom. We therefore test $F = V_b/V_r$ against critical F with $v_1 = m - 1$, and $v_2 = N - m$. If this observed F is significantly large we estimate σ_b^2 from

$$\frac{N(m - 1)(V_b - V_r)}{N^2 - (n_1^2 + \ldots + n^2 m)} \to \sigma_b^2 \quad (7.11)$$

(N.B. The symbol '$\to$' means 'estimates'.)

Example

7.4(i). Test σ_b^2 for significance and find estimates of σ^2, σ_b^2 when $n_1, n_2, \ldots n_6 = 45, 15, 14, 16, 15, 15$ and $\sum\sum x_{ij}^2 = 3,934$, and $S_1, S_2, \ldots S_6 = 56, 78, 75, 42, 64, -69$ (these are the coded results of observations on lengths of cuckoos' eggs found in nests of meadow pipit, tree pipit, sparrow, robin, wagtail, and wren).

7.5 Analysis of variance for two-way classified observations (two treatments)

Quite often we wish to know whether two different 'treatments' both have an effect and to measure the increases in variance due to them. For example, suppose we have a number m of machines spinning cotton thread and we wish to know whether there is any significant variation between the strength of the product from each machine, and, simultaneously, we wish to know whether a number n of different dyes also cause strength variation. If we apply the test of Section 7.3 separately to the product of each machine *and*

Table 7.3. TABLE OF TWO-WAY CLASSIFIED OBSERVATIONS

Levels of 1st treatment: 1   2   . . . m | Row sums

Levels of					Row sums
Levels of	1	x_{11}	x_{21} . . . x_{m1}		$S_{\cdot 1}$
2nd	2	x_{12}	x_{22} . . . x_{m2}		$S_{\cdot 2}$
Treatment	.	.	. . . . . .		.
	n	x_{1n}	x_{2n} . . . x_{mn}		$S_{\cdot n}$
Column sums	:	$S_{1\cdot}$	$S_{2\cdot}$ . . . . $S_{m\cdot}$		T = Grand total

Note: the x_{ij} may be 'coded' by subtracting A from each of them.

each dye, then since there are mn such products, at p observations per product we would make mnp observations in all.

However, if it is clear that the two 'treatments' do not 'interact' (e.g. when different machines comprise the first 'treatment' and the different dyes the second 'treatment', it is reasonable to suppose that though different machines produce slightly different threads the effect of the dyes on strength will be independent of the effect of machine, i.e. the two treatments do not 'interact') then we can simultaneously test for and estimate the effect (on variance) of both treatments using *only* mn observations.

We proceed as follows: Let there be m different levels of 1st treatment and n different levels of 2nd treatment. Set out the observations as in Table 7.3, and calculate the sums shown (N.B. $S_{1\cdot} + S_{2\cdot} + \ldots + S_{m\cdot} = T = S_{\cdot 1} + S_{\cdot 2} + \ldots + S_{\cdot n}$, this acts as a check).

The quantities shown in Table 7.4 are next calculated; here, σ_1^2

Table 7.4. ANALYSIS OF VARIANCE FOR TWO-WAY CLASSIFICATION

Source of variation	Sum of squares	Degrees of freedom	(S. of S)/ (D of F) estimates
Between columns	$\left(\sum\limits_{i=1}^{m} S_{i\cdot}^2 \right)/n - T^2/mn$	$m - 1$	$n\sigma_1^2 + \sigma^2$
Between rows	$\left(\sum\limits_{j=1}^{n} S_{\cdot j}^2 \right)/m - T^2/mn$	$n - 1$	$m\sigma_2^2 + \sigma^2$
Residual	$\sum\limits_{i=1}^{m} \sum\limits_{j=1}^{n} x_{ij}^2 + T^2/mn - \sum\limits_{i=1}^{m} S_{i\cdot}/n + \sum\limits_{j=1}^{n} S_{\cdot j}^2/m$	$(m-1)(n-1)$	σ^2
Total	$\Sigma\Sigma x_{ij}^2 - T^2/mn$	$mn - 1$	

is the true variance arising from the 1st treatment and σ_2^2 is that arising from the second treatment, and σ^2 is that due to all other causes (σ^2 is, therefore, usually called the 'residual' variance).

Writing

$V_1 = $ [Sum of squares between columns]$/(m - 1)$
$V_2 = $ [Sum of squares between rows]$/(n - 1)$
$V_r = $ [Residual Sum of Squares]$/(m - 1)(n - 1)$

we first test $F = V_1/V_r$ against critical F for $\nu_1 = m - 1$, $\nu_2 = (m - 1)(n - 1)$. If this observed F is significantly large we estimate σ_1^2 from

$$(V_1 - V_r)/n \rightarrow \sigma_1^2 \qquad (7.12)$$

Then we test $F = V_2/V_r$ against critical F for $\nu_1 = n - 1$, $\nu_2 = (m - 1)(n - 1)$ and, if this observed F is significantly large we estimate σ_2^2 from

$$(V_2 - V_r)/m \rightarrow \sigma_2^2 \qquad (7.13)$$

Note that

Between columns + Between rows + Residual sum of squares
= Total sum of squares

(the same relation holds for the degrees of freedom). We estimate σ^2 from

$$V_r \rightarrow \sigma^2$$

Since in Table 7.3 neither the observations in a single row nor those in a single column can be *rearranged*, the observations are said to be classified in 'two ways'.

Examples

7.5(i). If the Between columns, Between rows, Residual sum of squares = 5·2, 26·2, 4·5 and if $m = 4$, $n = 5$ test for the significance of σ_1^2, σ_2^2 and find estimates of σ^2, σ_2^2.

7.5(ii). Do as in Example 7.5(i) when the three sums of squares = 449, 106, 35; $m, n = 6, 24$.

Example of analysis of variance of two-way classified data

Suppose we have to analyse the data of Table 7.5. Carrying out the calculations of Table 7.4 we obtain

Sum of squares between columns

$$\begin{aligned}
&= (S_1.^2 + S_2.^2 + \ldots + S_5.^2)/n - T^2/mn \\
&= (144 + 225 + 576 + 729 + 144)/3 - 8{,}100/3 \times 5 \\
&= 606 - 540 \\
&= 66
\end{aligned}$$

Table 7.5. A Set of Two-way Classified Data

Levels of 1st treatment:	1	2	3	4	5 = m	Row sums
Levels 1	5	7	10	9	4	$S_{\cdot 1} = 35$
of 2nd 2	3	3	8	8	3	$S_{\cdot 2} = 25$
Treatment: 3 = n	4	5	6	10	5	$S_{\cdot 3} = 30$
Column sums	$S_{1\cdot} = 12$	$S_{2\cdot} = 15$	$S_{3\cdot} = 24$	$S_{4\cdot} = 27$	$S_{5\cdot} = 12$	$T = 90$

Dividing this by $m - 1 = 5 - 1 = 4$, we obtain as estimate of $n\sigma_1^2 + \sigma^2$

$$66/4 = 16{\cdot}5 \rightarrow n\sigma_1^2 + \sigma^2 = 3\sigma_1^2 + \sigma^2 \qquad (7.14)$$

Sum of squares between rows $= \dfrac{S_{\cdot 1}^2 + S_{\cdot 2}^2 + S_{\cdot 3}^2}{m} - \dfrac{T^2}{mn}$

$$= \left(\frac{1{,}225 + 625 + 900}{5}\right) - 540$$

$$= 550 - 540 = 10$$

Dividing this by $n - 1 = 3 - 1 = 2$, we obtain the estimate

$$10/2 = 5 \rightarrow m\sigma_2^2 + \sigma^2 = 5\sigma_2^2 + \sigma^2 \qquad (7.15)$$

Total sum of squares
$$= \sum\sum x_{ij}^2 - T^2/mn$$
$$= (5^2 + 3^2 + 4^2 + 7^2 + \ldots + 4^2 + 3^2 + 5^2) - 540$$
$$= 628 - 540 = 88$$

Now

Residual

$= $ Total $-$ Between columns $-$ Between rows sums of squares
$= 88 - 66 - 10 = 12$

Dividing by $(m - 1)(n - 1) = 2 \times 4 = 8$, we obtain the estimate

$$12/8 = 1{\cdot}5 \rightarrow \sigma^2 \qquad (7.16)$$

If $\sigma_1^2 = 0$ then (7.14) and (7.16) are both estimates of σ^2. So to test the hypothesis that $\sigma_1^2 = 0$ we find the ratio of (7.14) to (7.16) and use the F test to see if this ratio is *greater* than would be expected from two different estimates of the same variance. Thus we obtain

$$16{\cdot}5/1{\cdot}5 = 11 > \text{crit } F(\nu_1 = m - 1, \nu_2 = \{m - 1\}\{n - 1\})$$
$$= F(\nu_1 = 4, \nu_2 = 8) = 7{\cdot}01$$

at the 1% level. So we confidently reject the hypothesis that $\sigma_1^2 = 0$ and we estimate σ_1^2 from

$$16\cdot5 - 1\cdot5 \to n\sigma_1^2 + \sigma^2 - \sigma^2 = n\sigma_1^2 = 3\sigma_1^2$$

i.e.

$$5 \to \sigma_1^2 \tag{7.17}$$

However, when we test the hypothesis that $\sigma_2^2 = 0$, the ratio of (7.15) to (7.16) is

$$5/1\cdot5 = 3\cdot33$$

$$F(\nu_1 = n - 1, \nu_2 = m - 1, n - 1) = F(\nu_1 = 2, \nu_2 = 8) = 4\cdot46$$

at the 5% level. Since 3·33 is not much lower than 4·46 the decision whether to accept that $\sigma_2^2 = 0$ or not is usually based on further considerations (e.g. past experience might have shown that σ_2^2 is small but not negligible, though in such borderline cases it is better to consult a professional statistician). If we decide not to neglect σ_2^2 then our best estimate is

$$5 - 1\cdot5 \to m\sigma_2^2 + \sigma^2 - \sigma^2 = 5\sigma_2^2$$

i.e.

$$0\cdot7 \to \sigma_2^2 \tag{7.18}$$

If, however, we decide to neglect σ_2^2 then (7.15) gives 5 as an estimate of σ^2 with 2 degrees of freedom and (7.16) gives 1·5 as an estimate of σ^2 with 8 degrees of freedom. We combine them to get the estimate

$$(5 \times 2 + 1\cdot5 \times 8)/(2 + 8) = 2\cdot2 \to \sigma^2 \tag{7.19}$$

with $2 + 8 = 10$ degrees of freedom. The ratio of (7.13) to (7.18) is

$$16\cdot5/2\cdot2 = 7\cdot5 > F(\nu_1 = m - 1 = 4, \nu_2 = 10) = 5\cdot99$$

at the 1% level; so we are still confident that σ_1^2 is not zero. A revised estimate of σ_1^2 is

$$16\cdot5 - 2\cdot2 = 14\cdot3 \to n\sigma_1^2 + \sigma^2 - \sigma^2 = 3\sigma_1^2$$

i.e.

$$4\cdot77 \to \sigma_1^2 \tag{7.20}$$

which is not much different from the estimate 5 of (7.17).

Examples

7.5(iii). Analyse the observations of Example 7.2(iii) as if they were two-way classified.

7.5(iv). Do as for Example 7.5(iii) for Example 7.2(iv).

7.6 More than two treatments, design of experiments

Analysis of variance can be extended to three or more treatments if they do not interact and if there are the same number of levels in each treatment (or the experiment can be arranged in parts in which this is true). If so, then only n^2 observations are needed where n is the number of levels in any one treatment. The carrying out of the observations has to be done in a special way described in books on the 'design of experiments' which should be consulted, or a professional statistician's services should be employed.

8

Normal sampling theory: estimation of 'parameters' by confidence intervals, by maximum likelihood

8.1 Definition of parameter

We have seen in Section 4.2 that a normal population is completely specified if its mean μ and standard deviation σ are known. For this reason μ and σ are often called 'parameters' of the normal distribution. In general, if the values of certain properties of a theoretical distribution (e.g. its mean, variance, 3rd moment, etc.) are known and if the distribution is then completely specified (i.e. if its frequency curve can be plotted) these properties are called 'parameters' of the distribution. The number of such properties required to be known before a distribution is completely specified is called 'the number of parameters of the distribution'. Thus the normal distribution has two parameters. But these are not necessarily μ and σ, and, strictly speaking, any two properties that will specify a normal distribution are entitled to be called its parameters (for example, if we know the values of $\mu + \sigma$ and $\mu - \sigma$, the distribution is specified, because the values of μ and σ would in fact, though not in appearance, be known; again, μ and v_3, see Section 4.4, can act as parameters).

The term 'parameter' is, however, usually restricted to those quantities appearing as letters (apart from the variate, e.g. x) in the formula for the probability density function (defined in Section 3.2). Thus, for the normal distribution, the probability density function is

$$\frac{1}{\sigma\sqrt{(2\pi)}} \exp\left[- \frac{(x - \mu)^2}{2\sigma^2} \right] \qquad (8.1)$$

and so μ and σ are usually what is meant by its 'parameters'.

Some theoretical distributions are specified completely by only one parameter (e.g. the value of μ specifies completely the Poisson distribution, as its probability formula (10.2) shows). Others require three or more (e.g. the normal bivariate of Chapter 13).

8.2 The nature of the estimation problem; confidence intervals

We now restrict our attention to the normal distribution and consider the problem of finding 'best' estimates for μ and σ from the observations in a sample. In Section 2.5 we stated that the mean $\bar{x}$ and observed variance s^2 are used as estimates of μ and σ^2. We showed in Section 5.6 that the expected values (i.e. averages over an infinite number of samples) of $\bar{x}$ and s^2 are μ and σ^2 (for this reason $\bar{x}$ and s^2 are called 'unbiased' estimates of μ and σ^2, see Section 5.6). However, there is no *a priori* reason why, say, the median observation of a sample (see Section 2.6) is not a better estimate of μ than $\bar{x}$; or, possibly, half the sum of the largest and smallest observations might be better (indeed, for the rectangular distribution it is better, see Section 10.5).

When the word 'best' is used in any context it has to be qualified before it becomes exact (e.g. teetotallers can agree that beer is 'best' —it is 'best left alone'). Now, in statistics there are several desirable properties which a best estimate should have; (i) it should be 'unbiased' (i.e. its expected value, see Section 3.5, should equal the true value of the parameter it is estimating), (ii) it should be a 'maximum likelihood' estimate (see Section 8.7), (iii) its variance (i.e. the expected value of (true parameter value − estimate)^2) should be a minimum (it is then called a 'most efficient' estimate, the reader can remember this term if he wishes).

In fact, $\bar{x}$ satisfies all these requirements and s^2 satisfies them with a certain reservation about (ii). But there is no reason why an estimate can always be found, whatever the distribution and whatever the parameter, which satisfies all those requirements simultaneously, and there are many cases where we know this is impossible.

Even if we accept $\bar{x}$ as the best estimate there is a further problem. We know that, in general, $\bar{x}$ is close to, but not *exactly* equal to μ. Consequently, we seek to specify a range or interval of values around $\bar{x}$ in which we can be confident that, very probably, μ lies. Such intervals are called 'confidence intervals' and we shall give formulae for them, both for μ and for σ^2.

(There are some alternative approaches to this problem; for example the 'inverse probability' approach which assigns relative 'probabilities' to values around $\bar{x}$. We shall not discuss them in this book.)

8.3   Confidence intervals for μ with σ known

We first consider the (relatively rare, but illustrative) problem in which we have a sample taken from a population whose standard deviation σ is known and where we desire to estimate μ. Now we know that the mean of a sample of size n is distributed normally with mean μ and standard deviation $\sigma/\sqrt{n}$ (see Section 5.7). By the properties of normal distributions (see Table 4.3), we know that for 95% of samples

$$\mu - 1 \cdot 96\sigma/\sqrt{n} < \bar{x} < \mu + 1 \cdot 96\sigma/\sqrt{n} \qquad (8.2)$$

Now, if $\mu - 1 \cdot 96\sigma/\sqrt{n} < \bar{x}$ then $\mu < \bar{x} + 1 \cdot 96\sigma/\sqrt{n}$. Again, if $\bar{x} < \mu + 1 \cdot 96\sigma/\sqrt{n}$ then $\bar{x} - 1 \cdot 96\sigma/\sqrt{n} < \mu$. Hence we can say that for 95% of samples

$$\bar{x} - 1 \cdot 96\sigma/\sqrt{n} < \mu < \bar{x} + 1 \cdot 96\sigma/\sqrt{n} \qquad (8.3)$$

and we call $\bar{x} - 1 \cdot 96\sigma/\sqrt{n}$ the 'lower 95% confidence limit' for μ, and $\bar{x} + 1 \cdot 96\sigma/\sqrt{n}$ the 'upper 95% confidence limit, for μ. The interval between these limits is called the '95% confidence interval' for μ.

Thus, if $\bar{x} = 14 \cdot 53$, $\sigma = 2 \cdot 0$, $n = 16$ then $1 \cdot 96\sigma/\sqrt{n} = 0 \cdot 98$ and the 95% confidence limits for μ are $14 \cdot 53 - 0 \cdot 98 = 13 \cdot 55$ and $14 \cdot 53 + 0 \cdot 98 = 15 \cdot 51$.

By a similar argument (see Table 4.3)

$$\bar{x} - 2 \cdot 58\sigma/\sqrt{n} < \mu < \bar{x} + 2 \cdot 58\sigma/\sqrt{n} \qquad (8.4)$$

gives the 99% confidence limits and interval for μ; and if we replace $2 \cdot 58$ by $3 \cdot 29$ in (8.3), we obtain the 99·9% confidence limits.

Example

8.3(i). Calculate the 95, 99, 99·9% confidence limits for μ when

$$\bar{x}, \sigma, n = (a) \ 7 \cdot 35, \ 1 \cdot 97, \ 10$$
$$(b) \ -4 \cdot 83, \ 20 \cdot 35, \ 6$$
$$(c) \ 81 \cdot 4, \ 10 \cdot 9, \ 49.$$

In fact, if, having obtained the value of $\bar{x}$ from a sample, we say that μ lies within the 95% interval (8.3) (or the 99% (8.4)) we shall make a mistake on $5 = 100 - 95$ (or on $1 = 100 - 99$)% occasions *whatever* the true value of μ. (The complete proof of this statement will not be given here.)

8.4 Confidence intervals for μ (σ unknown); relation to the *t* test

In most practical problems σ is unknown and we use s instead, where

$$s^2 = \{(x_1 - \bar{x})^2 + \ldots + (x_n - \bar{x})^2\}/(n - 1)$$

is the usual estimate of σ^2 (see (2.10)). Now, in Chapter 6 when σ^2 was replaced by s^2, as in Section 6.5, the test-function c (with critical values $1 \cdot 96$, $2 \cdot 58$, $3 \cdot 29$) was replaced by the *t* function (critical values given by Table 6.2 or Table A.3). Similarly here the confidence interval for μ becomes, by analogy with 8.2,

$$\bar{x} - t_c s/\sqrt{n} < \mu < \bar{x} + t_c s/\sqrt{n} \qquad (8.5)$$

where t_c is the critical value of t, at the desired probability level, for degrees of freedom $\nu = n - 1$. To get the 95, 99, or 99·9% confidence interval the probability level for t_c must be the 5, 1, or 0·1% level, respectively.

Suppose $\bar{x} = 29\cdot46$, $n = 9$, $s = 1\cdot44$. Now $\nu = 8$ and from Table A.3 we see that at the 5% level the critical t value t_c is 2·31. Hence $t_c s/\sqrt{n} = 2\cdot31 \times 1\cdot2/3 = 0\cdot924$, and the confidence limits for μ are $29\cdot46 - 0\cdot924 = 28\cdot54$ and $29\cdot46 + 0\cdot924 = 30\cdot38$ (we only work to 2 decimal places, since the original figure for $\bar{x}$ is only correct to this level).

Examples

8.4(i). Calculate the 95, 99, 99·9% confidence intervals for μ given

$$\bar{x}, s, n = (a) \quad 17\cdot4, 11\cdot8, 10$$
$$(b) \quad -21\cdot6, 12\cdot8, 7$$
$$(c) \quad -5\cdot07, 4\cdot46, 24.$$

8.4(ii). Calculate the 95, 99, 99·9% confidence intervals for μ given the sample of observations 12·32, 10·95, 4·98, 14·03, 8·97, 5·83.

8.4(iii). Calculate the 95, 99, 99·9% confidence intervals for μ given the sample $-4\cdot8$, 11·2, $-3\cdot1$, 2·1, 8·7.

The t test of Section 6.5 has a simple relation with the confidence interval (8.5). In fact, the difference between the sample mean $\bar{x}$ and a given value μ' for μ is said to be significant at the 5, 1, or 0·1% level if μ' lies *outside* the 95(= 100 − 5), 99 (= 100 − 1), or 99·9(= 100 − 0·1)% confidence interval (given by (8.5) with the appropriate value of t_c). Thus if we wish to test the hypothesis that the population mean is 32·00 and (see example above) we have $\bar{x} = 29\cdot46$, $n = 9$, $s^2 = 1\cdot44$; then since 32·00 lies outside the 95% confidence interval 28·54, 30·38 we reject the hypothesis at the 5% level.

Example

8.4(iv). With the above data, can we reject the hypothesis that the mean is: (a) 32·00, (b) 27·28 at the 1, and at the 0·1% level?

8.5 Confidence intervals for the difference between two means

(a) **Case of equal population variances.** Let $\bar{x}$ and $\bar{x}'$ be the means of two samples and n, n' their sizes, and let S^2 be the combined estimate of their common population variance σ^2 as calculated by formula (6.9). The 95, 99, or 99·9% confidence limits for the difference $\mu - \mu'$, say, between the means μ and μ' of the populations from which the samples come, are

$$(\bar{x} - \bar{x}') - t_c S \sqrt{\left(\frac{1}{n} + \frac{1}{n'}\right)} < \mu - \mu' < (\bar{x} - \bar{x}') + t_c S \sqrt{\left(\frac{1}{n} + \frac{1}{n'}\right)}$$

(8.6)

where t_c is the critical value of t at the 5, 1, or 0·1% level (with degrees of freedom $v = n + n' - 2$).

Note that: (i) we do not estimate μ and μ' separately, (ii) the confidence interval (8.6) and the significance test of (6.11) are related in the following way: if the confidence interval is *entirely* positive or *entirely* negative (in other words, if it does not include *zero*) then we are confident that $\mu - \mu'$ could not be zero, and, hence, the samples are unlikely to come from populations with the *same* (true) mean.

Example

8.5(i). Calculate, assuming population variances to be the same, the 95, 99, 99·9% confidence intervals for $\mu - \mu'$ for the data of Examples 6.8(i) and (ii).

(*b*) **The general case (population variances in any ratio).** Let $\bar{x}$, $\bar{x}$; n, n'; s^2, $(s')^2$ be the sample means, sizes and variances. The confidence intervals here for the difference $\mu - \mu'$ between the population means is

$$(\bar{x} - \bar{x}') - W_c \sqrt{\left(\frac{s^2}{n} + \frac{(s')^2}{n'}\right)} < \mu - \mu' < (\bar{x} - \bar{x}')$$
$$+ W_c \sqrt{\left(\frac{s^2}{n} + \frac{(s')^2}{n'}\right)} \quad (8.7)$$

where W_c is the critical value of W at the desired probability level (see Table A.4).

Hence, if $\bar{x}$, $\bar{x}'$, n, n', s^2, $(s')^2 = 453, 473, 9, 11, 540, 440$ then (see Table A.4) W_c at the 5% level, for $n - 1 = 8$, $n' - 1 = 10$, $h = 0·6$ is 2·08 and $\sqrt{\{s^2/n + (s')^2/n'\}} = 10$, so we have $-40·8 < \mu - \mu'$ $< + 0·8$. It is clear from this interval that $\mu - \mu'$ could be zero and so we cannot be confident that $\mu \neq \mu'$ (see the significance test of Section 6.8(*b*)).

Example

8.5(ii). Calculate the 95% confidence interval for $\mu - \mu'$ for the data of Example 6.8(i) (*c*), (*d*) (note that with these we know that we shall make a mistake on only 5 per cent of occasions, but with the results of Example 8.5(i) we are not sure, unless the population variances *really* are equal.)

8.6 Confidence interval for σ^2

(a) **The χ^2-distribution.** Mathematicians have worked out the sampling distribution of the function (or 'statistic', to use the modern term, see Section 5.1)

$$\chi^2 \equiv \{(x_1 - \bar{x})^2 + (x_2 - \bar{x})^2 + \ldots + (x_n - \bar{x})^2\}/\sigma^2 \quad (8.8)$$

where, as usual, $x_1, \ldots x_n$ denote the observations in a sample of size n from a normal population with variance σ^2. (This distribution, the χ^2-distribution, as it is called, is the *same* whatever the exact values of μ and σ.) Note that

$$\chi^2 \equiv (n - 1)s^2/\sigma^2 \quad (8.9)$$

where s^2 is the sample variance (see Section 5.2). (The Greek letter χ is pronounced 'kigh', though it is spelt 'chi'. χ^2 can be pronounced as 'kigh-square' or 'kigh-squared' and is listed in the index under C as 'chi-square(d)'.) The reader should regard χ^2 as a single letter, denoting a variate with lowest possible value 0 and highest possible value $+\infty$; this is the reason for using χ^2 and not χ, because a square cannot be negative and a look at (8.8) or (8.9) will show that *whatever* $x_1, \ldots x_n$, χ^2 must be positive.

The sampling distribution (i.e. the probability density with which different values occur) of χ^2 is unlike most distributions mentioned previously in that it is not symmetric, see Fig. 8.1. This lack of symmetry means that when we find critical values we have to specify an upper critical value (for the upper or right-hand tail, see Fig. 8.1) and a lower critical value (for the left-hand tail). (In the case of the c and t functions of Chapter 6 the lower critical value is minus the upper critical value; there is no simple relation between the χ^2 critical values.) These values are denoted thus: $\chi^2_{0.025}$, see Fig. 8.1, denotes the critical value which is such that 2.5% of random samples give a higher value for χ^2 (this can be written $\Pr(\chi^2 > \chi^2_{0.025}) = 0.025$); similarly $\chi^2_{0.975}$ is such that 97.5% of samples give a higher value (i.e. $\Pr(\chi^2 > \chi^2_{0.975}) = 0.975$), in the latter case we see that $\Pr(\chi^2 < \chi^2_{0.975}) = 0.025$. Similarly, it follows that

$$\Pr(\chi^2 > \chi^2_{0.01}) = 0.01, \quad \Pr(\chi^2 < \chi^2_{0.99}) = 0.01, \text{ etc.}$$

Hence if s^2 is found from a sample there is a 95% probability that (see (8.9)),

$$\chi^2_{0.975} < (n - 1)s^2/\sigma^2 < \chi^2_{0.025} \quad (8.10)$$

and a 99% probability that

$$\chi^2_{0.995} < (n-1)s/\sigma^2 < \chi^2_{0.005} \qquad (8.11)$$

Table A.5 gives critical χ^2 values. Note that the critical values vary with what is called the 'degrees of freedom' ν.

In this section $\nu = n - 1$. In other applications ν may be given by a different formula. (N.B. The 99.9% probability level critical values $\chi^2_{0.999}$ are rarely used in practice.)

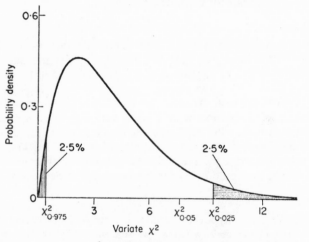

FIGURE 8.1. The χ^2 distribution curve for degrees of freedom $\nu = 3$, showing tails and critical values

(b) **Confidence intervals for σ^2.** If s^2 is calculated from a random sample then the 95% confidence interval for σ^2 is given by

$$\frac{(n-1)s^2}{\chi^2_{0.025}(\nu = n-1)} < \sigma^2 < \frac{(n-1)s^2}{\chi^2_{0.975}(\nu = n-1)} \qquad (8.12)$$

and we would be wrong on $5 = 100 - 95\%$ of occasions if we say that σ^2 lies in this interval. Note that when we wish to mention the degrees of freedom explicitly we write $\chi^2(\nu = n-1)$, instead of just χ^2. The 99% interval is

$$(n-1)s^2/\chi^2_{0.005}(\nu = n-1) < \sigma^2 < (n-1)s^2/\chi^2_{0.995}(\nu = n-1)$$
$$(8.13)$$

where as usual

$$s^2 = \{(x_1 - \bar{x})^2 + \ldots + (x_n - \bar{x})^2\}/(n-1)$$

Thus suppose n, $s^2 = 9$, 12·3 then $\nu = 9 - 1 = 8$ and

$$\chi^2_{0\cdot975}(\nu = 8) = 2\cdot18, \quad \chi^2_{0\cdot025}(\nu = 8) = 17\cdot5.$$

Hence we can say that the 5% confidence interval is

$$8 \times 12\cdot3/17\cdot5 = 5\cdot62 < \sigma^2 < 8 \times 12\cdot3/2\cdot18 = 45\cdot14$$

The interval is thus quite large, the upper limit 45·14 being over 8 times the lower limit 5·62. This is, unfortunately, a feature of confidence intervals for σ^2 (unless n is large).

Example

8.6(i). Calculate 95 and 99% confidence intervals for σ^2 when s^2, $n = $ (a) 10·7, 6; (b) 94·5, 11; (c) 0·0658, 16; for the data of (d) 8.4(ii), (e) 8.4(iii), (f) the two samples of Example 5.6(i).

(c) **Significance test.** To test the hypothesis that the sample could come from a population with a given variance value σ_0^2, say, we calculate the 95 (or the 99)% confidence interval (8.12). If σ_0^2 lies *outside* this interval we reject the hypothesis at the 5 (or 1)% probability level (for, if the sample really did come from a population with variance σ_0^2, we would be wrong on 5 (or 1)% of occasions).

Example

8.6(ii). Test the hypothesis that: (a) $\sigma_0^2 = 14$ when $s^2 = 10\cdot7$, $n = 6$; (b) $\sigma_0^2 = 68\cdot7$ when $s^2 = 94\cdot5$, $n = 11$; (c) $\sigma_0^2 = 0\cdot046$ when $s^2 = 0\cdot0658$, $n = 16$; (d) $\sigma_0^2 = 193\cdot2$ when $s^2 = 101\cdot5$, $n = 100$.

8.7 Estimates of μ and σ^2 by maximum likelihood

This section describes the principle of estimating by 'maximum likelihood'. The reader can omit it and accept $\bar{x}$, s^2 as best estimates if he wishes.

Now the probability density function for a normal distribution is, see equation (4.3)

$$\frac{1}{\sigma\sqrt{(2\pi)}} \exp\left[-\frac{(x - \mu)^2}{2\sigma^2}\right] = L(x) \qquad (8.14)$$

Suppose we just had one observation $x = 5$, say. Then

$$L(5) = \frac{1}{\sigma\sqrt{(2\pi)}} \exp\left[-\frac{(5 - \mu)^2}{2\sigma^2}\right] \qquad (8.15)$$

is called a 'likelihood' function, and the values of μ and σ which make $L(5)$ a maximum are the 'maximum likelihood' values for μ

and σ. Suppose we try $\mu = 0$ and $\sigma = 10$, then $L(5) = 0·035$; but if we try $\mu = 5$ and $\sigma = 1$, then $L(5) = 0·399$, and we feel that the latter values are much more likely to be the true population values.

Extending this idea to more than one observation, we multiply the likelihood function for each observation to get the combined likelihood function. (We multiply because the probability of getting observations, $x_1, x_2, \ldots x_n = \Pr(x_1)\Pr(x_2) \ldots \Pr(x_n)$, see the laws of probability, Section 9.2.) That is, if $x_1, \ldots x_n$ are the observations, then

$$L(x_1)L(x_2) \ldots L(x_n) \tag{8.16}$$

is the combined likelihood function, and the values of μ and σ which make this a maximum are the 'maximum likelihood estimates'. By the usual calculus methods for finding maxima, it can be proved that the maximum likelihood estimate for μ is

$$(x_1 + x_2 + \ldots + x_n)/n = \bar{x} \tag{8.17}$$

and the maximum likelihood estimate for σ^2 is

$$\{(x_1 + \bar{x})^2 + \ldots + (x_n - \bar{x})^2\}/n = (n - 1)s^2/n \tag{8.18}$$

and not s^2 itself, which may disappoint the reader. However, if μ is replaced by $\bar{x}$ in the formula (8.14) for $L(x)$ then the maximum likelihood estimate for σ^2 is s^2. This is preferred to (8.18) because, amongst other things, the expected value of (8.18) is $(n - 1)\sigma^2/n$ and not σ^2 (in other words the estimate (8.18) is biased).

In more complicated problems it sometimes proves very complicated or impossible to obtain confidence intervals and we have to be content with a single 'best' estimate based on some principle or other, and the method of maximum likelihood provides such a principle.

9

The binomial distribution: laws of probability, applications of the binomial distribution, the multinomial distribution

The binomial distribution is the most important of the non-normal distributions. Its most widely used application is estimating the 'fraction defective' in industry (the fraction defective is the proportion of articles which fail to meet a given standard, e.g. screws nominally of one inch length may be classed as defective if they are less than $\frac{63}{64}$ in. or greater than $\frac{65}{64}$ in. long). It is used in public opinion polls to estimate the proportion of the population who hold a certain opinion (e.g. who will vote for a certain political party), and has many other applications.

Probabilities in many card and other games are also associated with the binomial distribution (and with its generalization the multinomial distribution). Indeed, the origin of all probability theory was an investigation of a gambling problem which we now discuss.

9.1 Some problems in dice throwing

The aristocrat de Fermé, in 1654, asked two famous French mathematicians (Pascal and Fermat) to calculate the odds against three sixes appearing when a die is thrown three times.

They reasoned thus: If a die is unbiased, then each of the numbers 1, 2, 3, 4, 5, 6 should appear, in a long sequence of throws, an equal proportion of times (though not in any regular order). So, any given number, say a 4, will appear in $\frac{1}{6}$ of the throws. This is expressed in short by saying that 'the probability of a 4 is $\frac{1}{6}$'.

The next problem considered was that of obtaining two specified numbers when a die is thrown twice, e.g. a 4 on the first throw and a 5 on the second. Now, in a long sequence, a 4 will appear on $\frac{1}{6}$ of first and a 5 will appear on $\frac{1}{6}$ of second throws. So on $\frac{1}{6} \times \frac{1}{6}$ of double throws, a 4 will appear on the first throw and a 5 on the second. This is expressed in short by saying that 'the probability of a 4 on the first throw followed by a 5 on the second is $\frac{1}{6} \times \frac{1}{6}$'.

It can be seen that the probability of any specified number on the first throw followed by any other specified number on the second is also $\frac{1}{6} \times \frac{1}{6} = \frac{1}{36}$; e.g. the probability of a 2 followed by a 5 is $\frac{1}{36}$, as is the probability of two sixes.

A simple extension of this reasoning shows that, when a die is thrown three times, the probability of a 4 followed by a 5 followed by a 6 is $\frac{1}{6} \times \frac{1}{6} \times \frac{1}{6}$. Similarly, the probability of a 6 on all three throws is also $\frac{1}{6} \times \frac{1}{6} \times \frac{1}{6} = \frac{1}{216}$, thus solving de Fermé's problem.

Exactly the same reasoning applies in other cases such as the tossing of a coin. The probability of a head when an unbiased coin is tossed, is $\frac{1}{2}$. Hence the probability of two heads in two tosses is $\frac{1}{2} \times \frac{1}{2} = \frac{1}{4}$; similarly, the probability, in three tosses, of first a head, then a tail, then a head is $\frac{1}{2} \times \frac{1}{2} \times \frac{1}{2} = \frac{1}{8}$.

Another simple yet fundamental problem is to find the probability of obtaining *either* a 2 *or* 4, say when a single die is thrown. Obviously, in a long sequence of throws, in $\frac{1}{6}$ of them a 2 will appear, and in a different $\frac{1}{6}$ a 4 will appear. So the proportion of throws in which either a 2 or a 4 appears is $\frac{1}{6} + \frac{1}{6}$; in short 'the probability of either a 2 or a 4 is $\frac{1}{6} + \frac{1}{6}$'.

Note that the appearance of a 2 and of a 4 are said to be *mutually exclusive outcomes* of the throw of a die because a 2 and a 4 cannot appear at the same time. But in drawing a card from a pack the drawing of a spade and the drawing of a queen are not mutually exclusive outcomes as the queen of spades comes in both categories.

9.2 The laws of probability

The above ideas, extended to more general circumstances, form the 'laws' or 'rules' of probability. By means of these laws, probabilities in a wide range of complicated problems can be calculated clearly and simply.

The first law states that: if, in a 'trial' (such as the throw of a die, the toss of a coin, the selection of an individual from a population, etc.) a particular 'outcome' E (e.g. E could be the appearance of a 5, of a head, of an individual with a given variate-value, etc.,) has a probability p (i.e. occurs a proportion p of times); and, if in another trial (the toss of another coin, the selection of an individual from another population, etc.), the outcome E' has a probability p', then, when the two trials are held together, the probability (i.e. the proportion of times) that the outcomes E and E' occur *together* is $p \times p'$ (in technical language we say 'the *joint* probability of E and E' is pp''). Thus, if a coin is tossed and a die thrown the 'joint probability of a head *and* a 4' is $\frac{1}{2} \times \frac{1}{6} = \frac{1}{12}$.

In the succinct notation of Section 4.3 the first law states that

$$\text{Pr}(E \text{ and } E') = \text{Pr}(E) \times \text{Pr}(E') \qquad (9.1)$$

This law can be extended to several trials which take place together, thus:

$$\Pr(E \text{ and } E' \text{ and } E'' \text{ and} \ldots) = \Pr(E) \times \Pr(E') \times \Pr(E'') \times \ldots$$
(9.2)

Note that we assume the outcomes in each trial to take place at random, that is with no regularity or pattern whatsoever.

Examples

9.2(i). A coin is tossed and a die thrown, what is the joint probability of (a) a head and a 6, (b) a tail and either a 2 or a 5.

9.2(ii). A coin is tossed twice and a die thrown three times, what is the joint probability of: (a) two heads and three sixes, (b) two tails and either a 3 or a 4 on each throw.

The second law of probability says that if, in a trial the only possible outcomes are $E_1, E_2, \ldots E_k$ with probabilities $p_1, p_2, \ldots p_k$, then provided the outcomes are *mutually exclusive*, the probability of either E_1 or E_2 is $p_1 + p_2$; that is

$$\Pr(\text{either } E_1 \text{ or } E_2) = p_1 + p_2$$

For example the probability of drawing either the king of hearts or the queen of spades from a card-pack is $\frac{1}{52} + \frac{1}{52}$.

More generally, we have

$$\Pr(\text{either } E_1 \text{ or } E_2 \text{ or} \ldots \text{or } E_f) = p_1 + p_2 + \ldots + p_f \quad (9.3)$$

Thus the probability of drawing (a) a jack (of any suit) $= \frac{1}{52} + \frac{1}{52} + \frac{1}{52} + \frac{1}{52} = \frac{1}{13}$, (b) a spade is $\frac{13}{52}$, (c) either a spade or the ace of hearts $= \frac{13}{52} + \frac{1}{52} = \frac{14}{52}$.

But note that the outcomes of drawing a spade and of drawing a queen are not mutually exclusive as the queen of spades comes in both outcomes.

Note that if $E_1, E_2 \ldots E_k$ are mutually exclusive and are the *only possible* outcomes then

$$p_1 + p_2 + \ldots + p_k = 1 \quad (9.4)$$

for if $p_1 + \ldots + p_k < 1$ then on some proportion of occasions there would be some other outcome which is impossible.

9.2(iii). Find the probability of drawing from a 52 card-pack: (a) either a picture-card or a non-picture club; (b) either a black king or a diamond or a picture heart.

(Note that in many textbooks the word 'event' is used instead of 'outcome'; originally they had the same meaning but 'event' has nowadays acquired a wider range of meanings, so 'outcome' has been preferred here.)

For completeness we give the other two 'laws' or 'rules' of probability; they are really deductions from the first two (note too, that the order in which the laws appear varies from book to book).

Now, call drawing a spade from a pack outcome A; so drawing a non-spade (a card of another suit) is a Not-A outcome; and if we call drawing a picture-card outcome B, drawing a non-picture card will be Not-B. So we have four mutually exclusive categories for an outcome, namely, (i) both A and B (e.g. a picture-card spade), (ii) Not-A and B (a picture card of another suit), (iii) A and Not-B (a non-picture card spade), and (iv) Not-A and Not-B (a non-picture card of a non-spade suit). Now the third law concerns what is called 'conditional' probability; for example given that an outcome is B what is the probability (= proportion of occasions) that it is also A. This probability is written $\Pr(A|B)$ and called 'the conditional probability of A given B'; the word conditional is sometimes omitted. For example the probability that a drawn card is a king given that it is a heart, i.e. $\Pr(\text{King}|\text{Heart})$ is $1/13$.

Remembering that probability equals proportion, we see (after a little thought) that

$$\Pr(A|B) = \Pr(\text{both A and B})/\Pr(B) \qquad (9.5)$$

which is our third law of probability. Thus to find the conditional probability that, given that a drawn card is a picture-card (category B, say), it is also a spade (category A, say), we have to find $\Pr(\text{both A and B})$, i.e. the probability that a drawn card is both a picture-card and a spade; this is $\frac{3}{52}$. Now $\Pr(B) = \frac{12}{52}$, so, by (9.5),

$$\Pr(A|B) = (\tfrac{3}{52})/(\tfrac{12}{52}) = \tfrac{3}{12} = \tfrac{1}{4}.$$

9.2(iv). Find the conditional probability that, given that a card drawn from a 52-card pack: (a) is red, it is also a picture-card, (b) is a non-picture-card, it is also black.

The final law of probability concerns non-mutually exclusive outcomes, and states that

$$\Pr(\text{either A or B}) = \Pr(A) + \Pr(B) - \Pr(\text{both A and B}) \qquad (9.6)$$

The words 'either' and 'both' can be omitted, e.g. $\Pr(\text{both A and B})$ can be written $\Pr(A \text{ and } B)$. Thus, if we call being a spade category A, then $\Pr(A) = \frac{13}{52}$; and if we call being a queen category B, then $\Pr(B) = \frac{4}{52}$. Here the $\Pr(\text{both A and B})$ is the probability of being a queen and a spade and so is $\frac{1}{52}$. Hence the probability of a drawn card being either a spade or a queen is

$$\tfrac{13}{52} + \tfrac{4}{52} - \tfrac{1}{52} = \tfrac{16}{52} = \tfrac{4}{13}.$$

The extension to three non-mutually exclusive outcomes is

$$Pr(A \text{ or } B \text{ or } C) = Pr(A) + Pr(B) + Pr(C) - Pr(A \text{ and } B)$$
$$- Pr(A \text{ and } C) - Pr(B \text{ and } C) + Pr(A \text{ and } B \text{ and } C) \quad (9.7)$$

9.2(v). Find the probability that a drawn card is either a king, or a red picture-card or a spade.

9.3 The probability of two sixes in three throws of a die, etc.

By 'two sixes in three throws' we mean that a 6 appears in two throws only, it does not matter which two, but the other throw must *not* give a six.

We now apply the first two laws of probability to solve this problem, by finding separately the probability (i) of 6, 6, x occurring, where x means Not-6, (ii) of 6, x, 6 occurring, and (iii) of x, 6, 6. Now, by the first law, the probability of (i) is

$$\tfrac{1}{6} \times \tfrac{1}{6} \times \tfrac{5}{6} = \tfrac{5}{216}$$

that of (ii) is

$$\tfrac{1}{6} \times \tfrac{5}{6} \times \tfrac{1}{6} = \tfrac{5}{216}$$

and that of (iii) is

$$\tfrac{5}{6} \times \tfrac{1}{6} \times \tfrac{1}{6} = \tfrac{5}{216}$$

also. Now if we regard three throws as a *single* trial, the outcomes (i) and (ii) are mutually exclusive, since, if 6 occurs on the second throw, x cannot. Similarly, (i) and (iii), and (ii) and (iii), are mutually exclusive. So, by the second law, the probability of *either* (i) *or* (ii) *or* (iii) is

$$\tfrac{5}{216} + \tfrac{5}{216} + \tfrac{5}{216} = 3 \times \tfrac{5}{216} = \tfrac{5}{72} \quad (9.8)$$

Since (i), (ii) and (iii) cover all cases in which two sixes only appear (9.8) gives the answer to our problem.

Now consider the probability of two sixes in five throws. This can happen in a number of mutually exclusive ways, e.g. 6, 6, x, x, x; 6, x, 6, x, x; x, 6, 6, x, x, etc.; in fact in $^5C_2$ ways (see Note on Algebra, Appendix). The probability of 6, 6, x, x, x is

$$\tfrac{1}{6} \times \tfrac{1}{6} \times \tfrac{5}{6} \times \tfrac{5}{6} \times \tfrac{5}{6} = (\tfrac{1}{6})^2 (\tfrac{5}{6})^3,$$

which is also the probability of 6, x, 6, x, x; of x, 6, 6, x, x, etc. So the probability of *either* 6, 6, x, x, x or any other of the $^5C_2$ ways is

$$^5C_2 (\tfrac{1}{6})^2 (\tfrac{5}{6})^3 = (5 \times 4/1 \times 2)(\tfrac{1}{36})(\tfrac{5}{216}) = 1{,}250/7{,}776 = 0 \cdot 1607 \quad (9.9)$$

Examples

9.3(i). Find the probability: (*a*) of three sixes in five throws, (*b*) of four sixes in seven throws of a die.

9.3(ii). In two throws of a die find the probability that the sum is: (a) 5, (b) 10, (c) less than 5. (Hint. List the ways in which 5 can arise, i.e. 1, 4; 2, 3; 3, 2; 4, 1; and find the probability of each.)

9.4 The probability of s successes in n trials (the binomial probability)

The reasoning of Section 9.3 extends quite easily to the general case. If, in a single trial (e.g. a throw of a die) the probability of a 'success' is p (e.g. if a six is a success then $p = \frac{1}{6}$), then the probability of s successes in a set of n trials can be shown to be

$$^nC_sp^s(1-p)^{n-s} = \frac{n(n-1)(n-2)\ldots(n-s+1)}{1 \times 2 \times 3 \times \ldots \times n}p^s(1-p)^{n-s}$$
(9.10)

because (i) in n trials just s successes can appear in ^nC_s ways (see Note on Algebra, Appendix), (ii) the probability of occurrence of any single one of these ways, is, by the first law, $p^s(1-p)^{n-s}$, since the probability of non-success in a single trial is $1 - p$, (iii) the ways are mutually exclusive.

It is customary to write $q = 1 - p$, and (9.10) then becomes

$$^nC_sp^sq^{n-s}$$
(9.11)

This is the binomial probability formula and is well worth remembering. It is called 'binomial' because (9.11) is the term in p^s in the binomial expansion of $(p + q)^n$, see Note on Algebra, Appendix.

Consider this problem: Find the probability of two heads in five tosses of a coin. Here, p the probability of a success, i.e. a head, is $\frac{1}{2}$. Putting $n = 5$, $s = 2$, $p = \frac{1}{2}$ and, hence $q = 1 - \frac{1}{2} = \frac{1}{2}$, in (9.11) we get $^5C_2(\frac{1}{2})^2(\frac{1}{2})^3 = (5 \times 4/1 \times 2)(\frac{1}{2})^5 = \frac{5}{16}$.

Examples

9.4(i). Find the probability (a) of two tails in 6 tosses, (b) of 5 tails in 7 tosses of a coin.

9.4(ii). A card is drawn from a 52-card pack, then replaced and the pack shuffled before the next draw; find the probability of (a) 2 aces in 3 draws, (b) 3 picture-cards in 5 draws, (c) 4 hearts in 6 draws.

Another problem is: In a large consignment of articles (e.g. screws) one tenth are defective, what are the probabilities that, when 8 screws are selected at random, we find: (a) 0, (b) 1, (c) 2 defective? If we call finding a defective a 'success' then $p = \frac{1}{10}$. Putting $n = 8$, and $s = 0$ in (9.11), we find for: (a) the probability

$$^8C_0(\tfrac{1}{10})^0(\tfrac{9}{10})^8 = 1 \times 1 \times (0 \cdot 9)^8 = 0 \cdot 4305.$$

For (b) we put $s = 1$, and get

$$^8C_1(0 \cdot 1)^1(0 \cdot 9)^7 = 0 \cdot 3826$$

and, for (*c*) with $s = 2$, we get

$$^8C_2(0{\cdot}1)^2(0{\cdot}9)^6 = 15(0{\cdot}1)^2(0{\cdot}9)^6 = 0{\cdot}1488.$$

Note that there is a 43% probability of finding *no* defectives even though 10% of articles are defective. In testing for the fraction defective, in practice, often 100 or 200 or more articles have to be inspected to obtain an accurate value.

Example

9.4(iii). If the true fraction of articles defective is 0·2, find the probability that, out of 5 selected at random, 0, 1, 5 are defective.

9.5 The binomial distribution and some of its properties

The equivalence of 'probability' to 'proportion of times of occurrence' means that if we repeatedly took a set of n trials, then the proportion of times that a particular value of s occurred, $\Pr(s)$ say, would be given by formula (9.11). That is

$$\Pr(s) = {}^nC_s p^s q^{n-s}$$

Given values of n and p, $\Pr(s)$ can be calculated for each possible value of s, i.e. for $s = 0, 1, 2, \ldots n$. The set of values of $\Pr(s)$ forms a population or distribution (the variate is s) which is called a 'binomial distribution'.

Table 9.1 shows three calculated cases of binomial distributions. Note that in (i) and (ii) the probabilities are markedly asymmetric, but that (iii) shows some symmetry about its mode $s = 6$ (the mode is that value of s whose probability is a maximum). Also, (ii) is the 'mirror' image of (i); this is true for any two binomial distributions with the same n whose two values of p, p_1, and p_2 are such that $p_1 + p_2 = 1$; for then $q_1 = 1 - p_1 = p_2$, and $q_2 = 1 - p_2 = p_1$, and, since ${}^nC_s = {}^nC_{n-s}$ (see Note on Algebra, Appendix)

$${}^nC_s p_1{}^s q_1{}^{n-s} = {}^nC_{n-s} p_2{}^{n-s} q_2{}^s$$

or, in words, the probability of s successes in one distribution = probability of s non-successes (and therefore $n - s$ successes) in the other.

Table 9.1. VALUES OF $\Pr(s) = {}^nC_s p^s q^{n-s}$ FOR THREE BINOMIAL DISTRIBUTIONS

	n	p	s = 0	1	2	3	4	5	6	7	8	9	10	11	12	13	14	15
(i)	8	0·1	4305	3826	1488	0331	0046	0004	0000	0000	0000							
(ii)	8	0·9	0000	0000	0000	0004	0046	0331	1488	3826	4305							
(iii)	15	0·4	0005	0047	0219	0634	1268	1859	2066	1771	1181	0612	0245	0074	0016	0003	0000	0000

Now, all binomial distributions have certain properties in common. Thus

$$\sum_{s=0}^{n} \Pr(s) = \sum_{s=0}^{n} {}^nC_s p^s q^{n-s} = (p+q)^n = 1^n = 1 \qquad (9.12)$$

This result is to be expected from the general result (9.4) since the only possible outcomes $s = 0, 1, 2, \ldots n$ are mutually exclusive. It can be proved from the Binomial Theorem (see Note on Algebra, Appendix) because here $p + q = 1$.

The mean value, μ say, of s (averaged over an infinite number of sets of n trials) is

$$\mu = \sum_{s=0}^{n} s\Pr(s) = \sum_{s=0}^{n} s\,{}^nC_s p^s q^{n-s} = np(p+q)^{n-1} = np \qquad (9.13)$$

which might be expected intuitively, because the probability of success in one trial is p and so the average number of successes in n trials should be np.

The second moment ν_2 of s about the origin is by definition (see Section 3.4)

$$\nu_2 = \sum_{s=0}^{n} s^2\Pr(s) = \sum_{s=0}^{n} s({}^nC_s p^s q^{n-s}) = (n^2 - n)p^2 + np \qquad (9.14)$$

Examples

9.5(i). (For the mathematically-minded) Prove (9.13) by differentiating (9.12) partially with respect to p then putting $p + q = 1$.

9.5(ii). Prove (9.14) by differentiating (9.13) with respect to p.

The variance μ_2 of s is npq, a result worth remembering. This can be proved from the general formula $\mu_2 = \nu_2 - \mu^2$, see (2.7). Thus

$$\sigma^2 \equiv \mu_2 = (n^2 - n)p^2 + np - (np)^2 = np(1-p) = npq \qquad (9.15)$$

If np is 15 or more and not close to n, the binomial distribution is closely approximated by the normal distribution which has mean $\mu = np$, and variance $\sigma^2 = npq$; that is, instead of using (9.13) to calculate $\Pr(s)$, we put $s = x, \mu = np, \sigma = \sqrt{npq}$ and use the formula

$$\Pr(s) = p(x) \qquad (9.16)$$

where $p(x)$ is the normal probability density function of Table A.1. Thus, if $n = 100$ and $p = \frac{1}{5}$, we have $\mu = np = 20$, $\sigma = \sqrt{npq} = \sqrt{16} = 4$. If say, $s = 22$, then $|x - \mu|/\sigma = (22 - 20)/4 = 0\cdot5$; from Table A.1, the corresponding value of $\sigma p(x)$ is $0\cdot3521$, so by (9.16),

$$\Pr(s = 22) = 0\cdot3521/\sigma = 0\cdot0880.$$

9.5(iii). Calculate the probability that $s = 24, 18, 14$ for $n = 100$, $p = 1/5$.

Formula (9.16) is very much simpler than (9.11) for calculation, but, unfortunately, it is inaccurate if s is well away from np, and, though the probabilities are then small, in testing for the fraction defective we are concerned with just such cases.

The third and fourth moments of s about the mean, see Section 3.4, are

$$\mu_3 = npq(q - p), \quad \mu_4 = npq\{1 + 3(n - 2)pq\} \qquad (9.17)$$

9.5(iv). Calculate the mean and variance of s when: (a) $n = 6$, $p = 0.7$; (b) $n = 8, p = 0.15$; (c) $n = 100, p = 0.02$.

9.5(v). Calculate by (9.16) Pr(s) when $n = 10,000, p = 0.02$ and $s = 200, 207, 214, 193, 186, 179$.

If $n \geqslant 50$ and p is small but np is moderate, say between 0 and 10, the binomial distribution approximates to the Poisson (see Chapter 10).

The mode of s can be shown to be that integer ($=$ whole number) which is $\geqslant np - q$ and $\leqslant np + p$; if $np - q$ and $np + p$ are both integers then Pr($s = np - q$) and Pr($s = np + p$) are equal maximum probabilities; e.g. $n = 9, p = 0.4$.

9.6 Some observed binomial distributions

Since there is only a probability, not a certainty, that a seed of a given kind will germinate, an occasional packet may contain very few that germinate. It is important to know whether there are more of such packets than would be expected on probability theory.

To test whether the binomial distribution really applies in practice, an experiment was carried out in which 80 rows, each of 10 cabbage seeds, were examined after 10 days incubation. The results are shown in Table 9.2.

Table 9.2

Germinations in a row	0	1	2	3	4	5	6	7	$\geqslant 8$	Total
Observed frequency	6	20	28	12	8	6	0	0	0	80
Theoretical frequency	6.9	19.1	24.0	17.7	8.6	2.9	0.7	0.1	0.0	80

The theoretical frequency was calculated thus:
Total number of germinations

$$= (6 \times 0 + 20 \times 1 + \ldots + 8 \times 4 + 6 \times 5) = 174$$

So p the probability of a single seed germinating was put equal to $\frac{174}{800} = 0.2175$. With $n = 10$, $p = 0.2175$ then

$$\Pr(s) = {}^{10}C_s(0.2175)^s(0.7825)^{10-s},$$

and multiplying this by 80 gave the theoretical frequency for each value of s, i.e. for $s = 0, 1, \ldots 10$.

The agreement between observed and theoretical frequencies can be seen to be good (see Chapter 11 for a quantitative test).

Again, suppose we wish to test whether, among six-children families, the proportion of all-girl families is different from what would be expected on the hypothesis that the probability of a girl at a single birth is $\frac{1}{2}$. The theoretical probability of six 'successes' in six 'trials' is, by (9.11), ${}^6C_6(\frac{1}{2})^6 = 1 \times (\frac{1}{2})^6 = \frac{1}{64}$. The observed proportion is slightly lower than this (the true probability of a girl is, however, 0.488, and the correct proportion is, therefore,

$$ {}^6C_6(0.488)^6 = 0.0135 $$

which is in accord with observation).

9.7 A significance test

Given an observed number s_O of successes in a set of n trials, it is sometimes desired to test whether the probability p could have a given value p' (e.g. in 100 tosses of a coin we have 63 heads; could $p = \frac{1}{2}$ or is the coin biased?).

Obviously if s_O is too low or too high it will be unlikely that $p = p'$, but to determine what is too low or too high we must find the correct 'critical' values for s. We define s_l, the left-hand critical value of s, to be such that the probability of obtaining a value for s of 0, 1, 2 or . . . up to and including s_l, in a single set of n trials, is about 0.025, i.e. $\Pr(s \leqslant s_l) \simeq 0.025$, where

$$\Pr(s \leqslant s_l) = \Pr(0) + \Pr(1) + \Pr(2) + \ldots + \Pr(s_l).$$

Unfortunately, as s is not a continuous variate, we cannot make $\Pr(s \leqslant s_l) = 0.025$ *exactly*. Thus, if $n = 15$, $p = 0.4$ (see Table 9.1), $s_l = 2$ but

$$\Pr(s \leqslant 2) = 0.0005 + 0.0047 + 0.0217 = 0.0271.$$

If $n = 8$, $p = 0.1$, even if we put $s_l = 0$, the lowest possible value, $\Pr(s \leqslant 0) = 0.4305$. (However, in practical cases n is usually 50 or more and we can find s_l such that $\Pr(s \leqslant s_l) = 0.025$ fairly closely.) Similarly, the right-hand critical value s_r is such that $\Pr(s \geqslant s_r) = 0.025$. (Thus, s_l and s_r cut off the left- and right-hand tails of the binomial distribution.) Finally, if s_O is in one of these tails, i.e. if $s_O \leqslant$

s_l or $s_O \geqslant s_r$, we reject the hypothesis that $p = p'$, at about the 5% probability level (of making a Type I error, see Section 6.10). Similarly if $\Pr(s \leqslant s_l') \simeq 0.005$ and $\Pr(s \geqslant s_r') \simeq 0.005$, then if $s_O \leqslant s_l'$ or $s_O \geqslant s_r'$, we reject the hypothesis that $p = p'$ at the 1%, approximately, level.

If $np \geqslant 15$, to find s_l, s_r; s_l', s_r', we can use the normal approximation (9.16), and then, see Sections 4.3 or 6.2,

$$s_l = np - 1.96\sqrt{(npq)}, \qquad s_r = np + 1.96\sqrt{(npq)},$$
$$\tag{9.18}$$
$$s_l' = np - 2.58\sqrt{(npq)}, \qquad s_r' = np + 2.58\sqrt{(npq)}.$$

For example, if $n = 100$, $p = \frac{1}{2}$ then $np = 50$, $npq = 25$, and so $\sqrt{(npq)} = 5$; hence

$$s_r = 50 + 1.96 \times 5 = 59.8$$

and

$$s_r' = 50 + 12.9 = 62.9$$

So, if $s_O = 63$, we can reject the hypothesis that $p = \frac{1}{2}$ at the 1% level, and so strongly suspect that the coin is biased.

Examples

9.7(i). If, in 180 throws of a die (a) 22, (b) 41 sixes are recorded, test the hypothesis that p could equal $\frac{1}{6}$.

9.7(ii). In an extra sensory perception test, one man looked at a sequence of pictures of 5 different animals, another man who claimed telepathic powers, wrote down which of the five pictures he thought the first man was looking at. He obtained 26 successes out of 100. Test the hypothesis that $p = \frac{1}{5}$ (there is a probability of $\frac{1}{5}$ of being right at random; the one-tailed test applies here).

9.7(iii). (a) A diviner was invited to divine which of ten boxes covered a lump of metal. He was right twice in six tries. Test the hypothesis that $p = \frac{1}{10}$ (= the probability of being right at random, here).

(b) The above test is a two-tailed test, i.e. we reject if s_O is too low or too high. If, however, s_O is the number of defectives found among n articles from a consignment (or 'lot'), we are usually concerned with a one-tailed test, i.e. we wish to be confident only that p is not greater than some value p' (usually about $0.01 = 1\%$). If the observed value of s_O is so low as to make the hypothesis that $p = p'$ unlikely, we are in fact very pleased. In such cases, if $s_O \geqslant s_r$ we reject the hypothesis that $p = p'$ at the 2.5% level, for we shall make a mistake, if p really $= p'$, on only 2.5% of occasions.

9.8 Maximum likelihood estimates and confidence intervals for *p*

Often, having obtained s_O successes in n trials, we wish to estimate or find a confidence interval (see Chapter 8) for p (e.g. a Public Opinion Poll finds that 172 out of 1,000 people drink whisky regularly and a whisky distiller wants to know what limits can be assigned to p for the population as a whole).

(*a*) **Maximum likelihood estimate.** This estimate (see Section 8.7 for basic definition) is s_O/n, i.e. the observed proportion of successes.

(*b*) **Confidence limits (definition of).** The 95% limits are defined thus. The right-hand limit p_r is that value of p whose $s_l = s_O$, (where s_l is as defined in Section 9.7) and the left-hand limit is that value of p whose $s_r = s_O$.

(*c*) **Formula for confidence limits when $s_O \geqslant 15$; (testing hypotheses about *p*).** Here p_l, p_r are the roots of the following quadratic equation in p

$$p^2 \left[1 + \frac{(1\cdot96)^2}{n} \right] - \frac{p}{n} [2s_O + (1\cdot96)^2] + \frac{s_O^2}{n^2} = 0 \qquad (9.19)$$

If 1·96 is replaced by 2·58 the roots of 9·19 then give the 99% confidence limits for p.

There are prepared tables (which give slightly more accurate results than (9.19) which is based on the normal approximation to the binomial distribution), e.g. *Statistical Tables for Use with Binomial Samples* by Mainland, Herrera and Sutcliffe, New York, 1956.

Example

9.8(i). Find the 95 and 99% confidence limits for p given that $s_O, n = $ (*a*) 172, 1,000; (*b*) 30, 144; (*c*) 18, 200.

(*d*) **Estimating bird, fish populations, etc.** A number X, say, of the birds are captured, tagged and released. Later another number n, say, of the birds are recaptured and the number of these s_O who have tags is noted. The maximum likelihood estimate of the proportion of the whole bird population that is tagged is s_O/n. Since X birds were known to be tagged then nX/s_O is the maximum likelihood estimate of the bird population and X/p_l, X/p_r are the upper and lower confidence limits where p_l, p_r are the roots of (9.19).

Example

9.8(ii). Find the maximum likelihood estimate and 95, 99% confidence limits for population size when X, n, $s_O = $ (*a*) 500, 1,000, 45; (*b*) 120, 500, 28; (*c*) 3,000, 10,000, 196.

(e) **Confidence limits when $s_O < 15$, n large.** In testing for 'fraction defective' we examine a large number of articles (n is usually 50 or more and may be several hundreds) and p the true proportion defective is usually small so that s_O the observed number of defectives in the n articles is usually less than 15. In such cases we can use Table A.5 (which gives critical χ^2 values) to obtain very accurate confidence limits thus: Find $\chi^2_{0.005}$, $\chi^2_{0.025}$, $\chi^2_{0.975}$ and $\chi^2_{0.995}$ for degrees of freedom $\nu = 2s_O + 2$ from Table A.5, then: (a) the 95%, (b) the 99% confidence intervals for p are:

(a) $$\chi^2_{0.975}/2n < p < \chi^2_{0.025}/2n$$

(b) $$\chi^2_{0.995}/2n < p < \chi^2_{0.005}/2n \qquad (9.20)$$

However we usually, in fraction defective work, merely wish to be confident that p is just less than some stated value, say 0.02 (perhaps the manufacturer has promised the consumer that 98% of his articles are up to standard or perhaps the consumer has demanded this). In this case we use a one-tailed (one-sided is the better term) limit. We find, from Table A.5, $\chi^2_{0.050}$ and/or $\chi^2_{0.010}$ for $\nu = 2s_O + 2$ and we can be 95% confident that

$$p < \chi^2_{0.050}/2n \qquad (9.21)$$

or 99% confident that $p < \chi^2_{0.010}/2n$. So if we found that $\chi^2_{0.05}/2n > 0.02$ we could not be 95% confident that p, the true proportion defective, is less than 0.02; however, if $\chi^2_{0.05}/2n < 0.02$ we would pass the consignment as satisfactory.

For example suppose we find 4 defective out of 100 articles selected at random from a large consignment (or 'lot') then since, for $\nu = 8 + 2 = 10$, $\chi^2_{0.975} = 3.25$ and $\chi^2_{0.025} = 20.5$, the 95% two-tailed confidence limits for p are $3.25/200 = 0.01625$ and $20.5/200 = 0.1025$. Since, too, $\chi^2_{0.050} = 18.3$, for $\nu = 10$, we can be 95% confident that $p < 18.3/200 = 0.0915$.

Examples

9.8(iii). Find 95 and 99% (two-tailed) confidence limits given that $s_O, n = $ (a) 2, 50; (b) 3, 80; (c) 4, 100; (d) 9, 200; (e) 3, 300.

9.8(iv). Below what limit can we be 95, 99% confident that p lies for the data of Example 9.8(iii).

Quite often, in practice, we proceed this way: a number, say 50, of articles are selected from a 'lot' and (a) if very few, say 0 or 1, are found defective we pass the lot as satisfactory, (b) if many, say 10 or more are found defective we reject the lot, but (c) if between 1 and 10 are defective we examine a *further* 25 or 50 articles. This is called 'double sampling' or if the process is repeated several times 'sequential sampling'. It is not quite accurate to add the results of the two (or more) tests to get a combined s_O and a combined p. It is best to consult the tables to suit various circumstances which have been published, e.g. Dodge and Romig *Sampling Inspection Tables–Single and Double Sampling*, John Wiley & Sons, New York, or Columbia University Statistical Research Groups, *Sampling Inspection*, McGraw-Hill, New York.

Formulae (9.20) and (9.21) derive from the fact that the binomial distribution approaches the Poisson distribution (see Section 10.2) when n is large, p is small but $np < 15$.

The observant reader will note that in all the methods for estimating p we have nowhere mentioned the size of the population from which the n articles were selected except to say it is large. It is a remarkable feature of binomial estimation that it is almost completely independent of population size. Thus Public Opinion Polls question about 3,000 people to get a figure for the proportion of the British population who hold a given view. The same number would have to be interviewed to get the same accuracy in Monaco or India.

(*f*) **Testing hypotheses.** If we wish to test the hypothesis that p could have a certain value p' we calculate the 95 or 99% two-tailed confidence limits and if p' is outside these limits we reject the hypothesis at the $5(= 100 - 95)$ or $1 = (100 - 99)\%$ probability level.

Example

9.8(v). Test the hypothesis that (*a*) $p = 0.13$, (*b*) $p = 0.25$, for the data of 9.8(i) (*a*), (*b*) and (*c*).

9.9 Fraction defective when component parts are tested separately

(*a*) **2 component parts.** In many cases an article separates into two component parts which it is convenient or essential to test separately. We then need confidence limits for the fraction defective in the whole articles. The rule which follows has not been given before (though not exact it gives a good approximate answer).

Suppose the article splits into two parts and n_1 first parts are examined with s_1 found defective and n_2 second parts with s_2 found defective. We proceed thus: Find (by trial and error) λ' the positive value (which must also be less than $n_1 - s_1$ and $n_2 - s_2$) of λ which solves the equation

$$\lambda^2 \left[\frac{s_1}{n_1(n_1 - s_1 - \lambda)} + \frac{s_2}{n_2(n_2 - s_2 - \lambda)} \right] = 1.96^2 \qquad (9.22)$$

Then

$$1 - \frac{(n_1 - s_1 - \lambda')(n_2 - s_2 - \lambda')}{(n_1 - \lambda')(n_2 - \lambda')} \qquad (9.23)$$

gives the upper two-tailed 95% confidence limit for p the fraction defective in the articles as a whole.

Then we find the positive value λ'' which solves

$$\lambda^2 \left[\frac{s_1}{n_1(n_1 - s_1 + \lambda)} + \frac{s_2}{n_2(n_2 - s_2 + \lambda)} \right] = 1 \cdot 96^2 \qquad (9.24)$$

and

$$1 - \frac{(n_1 - s_1 + \lambda'')(n_2 - s_2 + \lambda'')}{(n_1 + \lambda'')(n_2 + \lambda'')} \qquad (9.25)$$

gives the lower two-tailed 95% confidence limit for p.

For example suppose $n_1, n_2, s_1, s_2 = 100, 120, 7, 5$ then $\lambda' = 43 \cdot 76$ and $\lambda'' = 78 \cdot 50$ and $0 \cdot 0634 < p < 0 \cdot 1818$ is the 95% confidence interval for fraction defective for the whole articles.

The trial and error finding of λ' and λ'' is quite simple; the left-hand sides of (9.22) and (9.24) are steadily increasing functions of λ.

If $1 \cdot 96$ is replaced by $2 \cdot 58$ we obtain the 99% two-tailed confidence limits and if by $1 \cdot 645$ or $2 \cdot 08$ then (9.23) gives the upper one-tailed (or one-sided) confidence limit.

(b) **k component parts.** The procedure is the same. Suppose we find n_1 first parts with s_1 defective, n_2 second parts with s_2 defective up to n_k final parts with s_k defective. Find positive λ' (less than $n_1 - s_1, n_2 - s_2, \ldots, n_k - s_k$) to solve

$$\lambda^2 \left[\frac{s_1^2}{n_1(n_1 - s_1 - \lambda)} + \cdots + \frac{s_k}{n_k(n_k - s_k - \lambda)} \right] = 1 \cdot 96^2 \qquad (9.26)$$

then

$$1 - \frac{(n_1 - s_1 + \lambda')(n_2 - s_2 - \lambda') \ldots (n_k - s_k - \lambda')}{(n_1 - \lambda')(n_2 - \lambda') \ldots (n_k - \lambda')} \qquad (9.27)$$

gives the upper two-tailed 95% confidence limit. Find positive λ'' to solve equation (9.28)

$$\lambda^2 \left[\frac{s_1}{n_1(n_1 - s_1 + \lambda)} + \cdots + \frac{s_k}{n_k(n_k - s_k + \lambda)} \right] = 1 \cdot 96^2 \qquad (9.28)$$

then

$$1 - \frac{(n_1 - s_1 + \lambda'') \ldots (n_k - s_k + \lambda'')}{(n_1 + \lambda'') \ldots (n_k + \lambda'')} \qquad (9.29)$$

gives the lower confidence limit. Again with $1 \cdot 96$ replaced by $2 \cdot 58$ we obtain the 99% two-tailed (or two-sided) confidence limits and

replaced by 1·645 or 2·33, (9.27) gives us the one-sided 95 or 99%
upper confidence limit.

Examples

9.9(i). Find 95 and 99% confidence limits for whole article
fraction defective given

$$s_1, n_1, s_2, n_2 = (a)\ 5, 100, 3, 80$$
$$(b)\ 2, 60, 7, 120$$
$$(c)\ 10, 90, 16, 150.$$

9.9(ii). Find 95 and 99% confidence limits for whole article
fraction defective given

$$s_1, n_1, s_2, n_2, s_3, n_3 = (a)\ 4, 90, 1, 80, 3, 120$$
$$(b)\ 5, 200, 4, 180, 7, 300.$$

9.9(iii). Find 95, 99% one-sided upper confidence limits for the
data of Examples 9.9(i) and (ii).

9.10 The multinomial distribution

This is a generalization of the binomial distribution. In the latter
we have either successes or non-successes, or defective or non-
defective articles, i.e. each article is in one of two mutually exclusive
classes. In some problems there are several mutually exclusive
classes, e.g. people can be blue-eyed, grey-eyed, brown-eyed, etc.

In such cases the multinomial distribution is appropriate. The
probability density formula is then as follows: If, in a large population, a proportion p_1 have characteristic C_1, p_2 have characteristic
$C_2, \ldots, p_k$ have C_k, then among n individuals selected at random
the probability that r_1 have characteristic C_1, r_2 have $C_2, \ldots, r_k$
have C_k is

$$\frac{n!}{r_1!\,r_2!\ldots r_k!}\,p_1^{r_1}p_2^{r_2}\ldots p_k^{r_k} \tag{9.30}$$

where $p_1 + p_2 + \ldots + p_k = 1$, $r_1 + r_2 + \ldots + r_k = n$ (the
characteristics $C_1, C_2 \ldots C_k$ are assumed to be mutually exclusive
and such that every individual has one and only one characteristic).
For example given that $\frac{1}{3}$ of people have fair hair, $\frac{1}{3}$ have brown,
$\frac{1}{4}$ have dark or black hair and the remainder red hair, what is the
probability that, out of 10 selected at random, 4, 3, 2, 1 have fair,
brown, dark, and red hair respectively. The answer is

$$\frac{10!}{4!\,3!\,2!\,1!}\left(\frac{1}{3}\right)^4\left(\frac{1}{3}\right)^3\left(\frac{1}{4}\right)^2\left(\frac{1}{12}\right) = \frac{7 \times 25}{24 \times 243} = 0{\cdot}030$$

Example

9.10(i). Find the probability that 3, 3, 2, 2, have fair, brown, dark and red hair, in the above case.

The probability density formula (9.30) can be proved in the same way as the binomial formula (9.11). We shall not discuss its properties except to say that if all the r_i are reasonably large (9.30) can be approximated by an extension of the normal approximation (9.16), and that this extension is the basis of the χ^2-test of Chapter 11.

10

The Poisson, the negative exponential, and the rectangular distributions

10.1 Typical Poisson distribution problems

These involve events which occur at random, with a certain probability. For example, cosmic ray particles arrive at random (at least to a very good approximation); if we know that, on the average, 10 particles arrive in a minute, then the number that arrive in any given hour will not necessarily be $10 \times 60 = 600$. So the probability distribution of the arrival of other numbers, e.g. 610 or 579, is of considerable importance.

Again, much queue theory is based on the assumption that customers arrive at random, and so involves this distribution and its properties.

It is named after the famous French mathematician Poisson (1781–1840).

10.2 The Poisson probability formula; some practical examples

If P is the probability that one random happening (e.g. the arrival of a cosmic ray particle from outer space) occurs in a small unit of time (say one second) then $\Pr(s)$, the probability of s happenings in a period of T units (e.g. s cosmic particles in one hour), can be proved to be

$$\Pr(s) = (PT)^s\, e^{-PT}/s! \qquad (10.1)$$

which is called the Poisson distribution formula. Note that: (a) $s! = 1 \times 2 \times 3 \times \ldots \times s$; (b) $e^{-PT} \equiv$ antilog $(-0{\cdot}4343PT)$, which is useful if negative exponential tables are not available; (c) s is a *discrete* variate whose possible variate values are 0, 1, 2, $\ldots \infty$.

Example

10.2(i). If the probability of a cosmic particle striking an apparatus in one second is (a) 0·02, (b) 0·005, calculate the probability of 0, 1, 2, 3, or 4 particles striking in 1 minute, 100 sec, 5 min.

It can be proved (see (10.3)) that if we observed the value of s for each of an infinite number of periods (of T units of time) the mean value of s would be PT. Now μ is usually used to denote the (true) mean value of a population so the Poisson probability formula (10.1) is more usually (and more concisely) written

$$\Pr(s) = \mu^s\, e^{-\mu}/s! \tag{10.2}$$

Note that μ is necessarily positive but may have *any* value between 0 and ∞, e.g. 0·617.

Table 10.1 gives a practical example of a Poisson distribution. Adams, in the *Journal of Civil Engineers*, November, 1936 gave the observations of Table 10.1 which recorded the number of vehicles

Table 10.1

Number in a period, s	0	1	2	3	>3
Observed frequency of s	94	63	21	2	0
Theoretical frequency of s	97·0	57·9	18·5	3·8	0·8

passing a point in each of 180 different 10-second periods. The theoretical frequency (third row of the table) was derived thus: A total of

$$0 \times 94 + 1 \times 63 + 2 \times 21 + 3 \times 3 = 111$$

vehicles were observed in a total of

$$94 + 63 + 31 + 2 = 180$$

different periods. The mean number of vehicles observed per period is thus $\frac{111}{180} = 0·617$. We therefore put $\mu = 0·617$ in formula (10.2). That is, the proportion of the 180 periods in which, theoretically, s vehicles were observed should be $(0·617)^s\, e^{-0·167}s!$ Multiplying this proportion by 180 gives the theoretical frequency (i.e. number of periods) in which s vehicles should have been observed. This calculation carried out for $s = 0, 1, 2, 3$ gives 97·0, 59·9, 16·5, and 3·8 and by subtracting the sum of these from 180 we obtained the theoretical frequency 0·8 for $s > 3$.

The observed and theoretical frequencies of Table 10.1 fit well together (but see Section 11.3 for a proper test for 'goodness of fit'), and such a good fit indicates that the passing of the cars was at random (i.e. there was little or no regularity). If a Poisson distribution fits such data well it is an indication that the happenings (arrival of cars, particles, customers, etc.) take place at random.

Example

10.2(ii). Fit Poisson distributions to each of the following three sets of data (given originally by Student).

Number of happenings in a period . .	0	1	2	3	4	5	6	7	8	9	10	11	12
Observed frequency (a)	103	143	98	42	8	4	2						
(b)	75	103	121	54	30	13	2	1	0	1			
(c)	0	20	43	53	86	70	54	37	18	10	5	3	2

10.3 Properties of the Poisson distribution

These may be remembered if desired.

(a) If we assume that the probability formula (10.1) holds, then the mean value of s, taken over an infinite number of periods, is

$$\sum_{s=0}^{\infty} s\mathrm{Pr}(s) = PT \sum_{s-1=0}^{\infty} (PT)^{s-1} \mathrm{e}^{-PT}/(s-1)! = PT\,\mathrm{e}^{-PT}\,\mathrm{e}^{+PT}$$

$$= PT \qquad (10.3)$$

(b) The most famous Poisson property is that the *variance* of s equals its *mean* (i.e. $\sigma^2 = \mu$). This can be proved thus. By virtue of (10.3) we can use (10.2) for $\mathrm{Pr}(s)$. Hence the variance of s is

$$\sum_{s=0}^{\infty} (s-\mu)^2\mathrm{Pr}(s) = \sum_{s=0}^{\infty} \{s(s-1)+s-2s\mu+\mu^2\}\mu^s\,\mathrm{e}^{-\mu}/s!$$

$$= \mu^2 + \mu - 2\mu^2 + \mu^2 = \mu \qquad (10.4)$$

(c) The third and fourth moments of s about its mean are

$$\sum(s-\mu)^3\mathrm{Pr}(s) = \mu, \quad \sum(s-\mu)^4\mathrm{Pr}(s) = \mu + 3\mu^2 \qquad (10.5)$$

(d) When n is large and p small, but $np < 15$ or thereabouts, the binomial probability formula (9.11) can be replaced by the Poisson formula (10.2) provided we put $\mu = np$. That is

$$^{n}\mathrm{C}_s p^s(1-p)^{n-s} \equiv n(n-1)\ldots(n-s+1)p^s(1-p)^{n-s}/s!$$
$$\simeq (np)^s\,\mathrm{e}^{-np}/s! \qquad (10.6)$$

This follows because, if n is large but s moderate,

$$n-1 \simeq n-2 \simeq \ldots \simeq n-s+1 \simeq n$$

and so

$$n(n-1)\ldots(n-s+1)p^s \simeq (np)^s.$$

Again, $(1-p)^n \simeq \mathrm{e}^{-np}$, which is a well-known result. By using the Poisson formula, binomial distribution probabilities are much more easily calculated.

Examples

10.3(i). If $n = 100$, $p = 0.03$ calculate $\Pr(s)$ using: (a) (9.11), (b) (10.6) for $s = 0, 1, 2, 3, 4$.

10.3(ii). (Mathematical) Prove the results of (10.5).

(e) There is a remarkable relation between the Poisson and χ^2-distributions (see Section 8.6), namely

$$\Pr(0) + \Pr(1) + \ldots + \Pr(s) = \alpha, \text{ if either } 2\mu = \chi^2_\alpha \text{ or } 2\mu = \chi^2_{1-\alpha} \tag{10.7}$$

where the critical χ^2, χ^2_α, and $\chi^2_{1-\alpha}$ both have degrees of freedom $\nu = 2s + 2$. (Properties (d) and (e) prove the formula (9.21) used in fraction defective testing.)

10.3(ii). Calculate $\Pr(0), \Pr(1), \ldots \Pr(s)$ for the following cases $\mu, s = $ (a) 0.82, 2; (b) 1.09, 3; (c) 1.08, 4 and show that one of the relations (10.7) holds in each case where for (a) $\alpha \simeq 0.95$, for (b) $\alpha \simeq 0.025$ and for (c) $\alpha \simeq 0.005$.

Tables of $\Pr(s)$ and $\Pr(0) + \Pr(1) + \ldots + \Pr(s)$ are available, e.g. *Handbook of Probability and Statistics* by Burington and May, Handbook Publishers, Sandusky, Ohio.

10.4   The negative exponential distribution

Whereas the Poisson distribution gives the probability of s happenings in a period of time if the happenings occur at random, the *intervals* of time between successive random happenings have a distribution of the type called 'negative exponential'. This type is such that if x denotes, as usual, the variate (which, as mentioned, could be the time between successive random happenings) the probability density $p(x)$ has the formula

$$p(x) = \beta\, e^{-\beta x} \tag{10.8}$$

where β can have any value between 0 and ∞. The least and greatest variate-values are $x = 0$ and $x = \infty$. The negative exponential distribution is a reverse J-shaped distribution (see Fig. 10.1) and the greater β the more sharply the distribution curve $p(x)$ slopes for $x \simeq 0$.

It has the following properties:

$$\mu = 1/\beta, \quad \sigma^2 = 1/\beta^2 \tag{10.9}$$

so that the standard deviation $= 1/\beta = $ the mean. These results can be proved from the formula for ν_r the rth moment about the origin, namely

$$\nu_r \equiv \int_{x=0}^{\infty} x^r \beta\, e^{-\beta}\, dx = r!/\beta^r \tag{10.10}$$

and the generally true relations $\nu_1 \equiv \mu$, $\nu_2{}^2 - \nu_1{}^2 \equiv \sigma^2$.

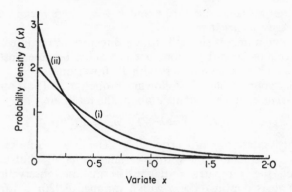

FIGURE 10.1. Two negative exponential distribution frequency
curves:

(i) $p(x) = 2\,e^{-2x}$, (ii) $p(x) = 3\,e^{-3x}$

The intervals between warp breaks in loom weaving have been
shown to have a distribution very close to the negative exponential;
it is a fair deduction, then, that warp breaks occur at random.

10.5 The rectangular distribution

This is not met with often in practice. Its frequency curve is a
rectangle, see Fig. 10.2, which means that its probability density
$p(x)$ is a constant. If the rectangle is of width a then $p(x)$ must be $1/a$,
because the area under a distribution curve *must* equal 1. The least

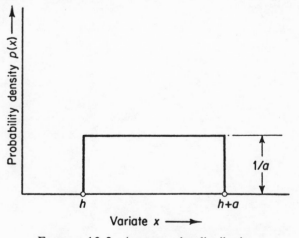

FIGURE 10.2. A rectangular distribution

and greatest variate-values are h and $h + a$, where h can have any value positive or negative.

Telephone-numbers selected from a directory at random have an almost rectangular distribution (but not quite, because Post Office engineers tend to keep numbers with 9's for testing).

The distribution has the following properties: The population mean, variance, and rth moment about the mean are

$$\mu = h + a/2, \quad \sigma^2 = a^2/12, \quad \mu_r = a^r/\{2^r(r + 1)\} \quad (10.11)$$

The most interesting property is this: The mean $\bar{x}$ of a sample of size n is an estimate of μ with a variance $a^2/(12n)$; but the 'mid point' of the sample ($= \frac{1}{2}\{$largest $+$ smallest observation$\}$) is also an unbiased estimate of μ with a variance $a^2/\{2(n + 1)(n + 2)\}$ which is *smaller* than $a^2/12n$ if $n > 2$ and very much smaller for large n.

11

The χ^2-test for 'goodness of fit' : test for 'association'

11.1 Nature of the test

Previously we have assumed that the correct type of distribution (e.g. normal, binomial, Poisson, etc.) for each problem was known. Obviously it is better to have some test available to determine quantitatively whether a set of observations really could come from a given (or a hypothetical) distribution. Karl Pearson in 1899 by using properties of the χ^2 distribution (see Section 8.6) devised the 'χ^2 test' for this purpose (there are other tests, but of a more specialized nature).

In its basic form it tests whether the observed proportions of individuals with given characteristics are significantly different from the proportions to be expected on some specified theory or hypothesis (it is therefore really a significance test for the multinomial distribution, see Section 9.10). But it is of wide application, for almost any other distribution can be arranged in a suitable form, see Section 11.8.

The test itself is quite simple and we now describe how to apply it.

11.2 The χ^2 test (applied to 'one-way' classified observations)

Suppose we wish to test the hypothesis that the proportions of people with different hair colours is as follows: $\frac{1}{3}$ fair, $\frac{1}{3}$ brown, $\frac{1}{4}$ dark or black, $\frac{1}{12}$ red or etc. Suppose, further, that the results of observations made on 240 people were.

Table 11.1

Hair colour . . .	Fair	Brown	Dark	Red
Observed frequency . .	72	87	68	13

(Note: Since individuals are here classed by hair-colour only, this is a 'one-way classification', classifying people by hair- *and* eye-colour, as in Section 11.4 gives a 'two-way classification'.)

Next, we find the frequency to be 'expected' in each class (of hair-colour, here) on the given hypothesis; that is, out of 240 people, we should expect $\frac{1}{3} \times 240 = 80$ to be fair, $\frac{1}{3} \times 240 = 80$ to be brown, $\frac{1}{4} \times 240 = 60$ to be dark, and $\frac{1}{12} \times 240 = 20$ to be red.

We now give the formula for *all* χ^2 tests, which is: Suppose $o_1, o_2, \ldots o_k$ are the observed and $e_1, e_2, \ldots e_k$ are the expected frequencies on a given hypothesis; calculate

$$\chi^2 = (o_1 - e_1)^2/e_1 + (o_2 - e_2)^2/e_2 + \ldots + (o_k - e_k)^2/e_k \quad (11.1)$$

then, if this value of χ^2 is *greater* than $\chi^2_{0.05}$, $\chi^2_{0.01}$, or $\chi^2_{0.001}$ (see Table A.5), we *reject* the hypothesis at the 5, 1, or 0.1 % probability level. The degrees of freedom ν of the critical χ^2 is given by the following rule:

ν = Number of classes − Number of relations between observed and expected frequencies $\quad (11.2)$

Thus, in our hair-colour example since there are 4 classes and one relation (the sum of the observed and of the expected frequencies are equal) then $\nu = 4 - 1 = 3$. Further,

$$\chi^2 = \frac{(72-80)^2}{80} + \frac{(87-80)^2}{80} + \frac{(68-60)^2}{60} + \frac{(13-20)^2}{20} = 4.93$$

which is *less* than $\chi^2_{0.05}$ ($\nu = 3$) = 7.81 (see Table A.5); so, we conclude that the observations are *not* inconsistent with the hypothesis of a $\frac{1}{3}$, $\frac{1}{3}$, $\frac{1}{4}$, $\frac{1}{12}$ distribution.

Examples

11.2(i). A die was thrown 1,200 times; there were 179 ones, 182 twos, 200 threes, 197 fours, 223 fives, 219 sixes. Test the hypothesis that the probability of appearance of each number is $\frac{1}{6}$.

Note that (i) we conclude there is a significant departure from hypothesis *only* if observed χ^2 is *greater* than critical (thus, it is a 'one-tailed' test, see Section 6.9; but see Example 11.2(ii) for a case of very small observed χ^2); (ii) the expected frequency in each class should be 20 or more for high accuracy, but, if not, Yates' correction, which we now describe, should be applied.

11.2(ii). In an experiment on crossing pea plants having round yellow seeds with others having wrinkled green seeds, the crossed plants had seeds as follows: round and yellow 305, wrinkled and yellow 101, round and green 103, wrinkled and green 35. Test the Mendelian hypothesis that the ratios are 9:3:3:1. Show that observed $\chi^2 < \chi^2_{0.99}$ (since such a low observed χ^2 would occur by chance *less than* 1 in a 100 times, we suspect that the experimenter has 'cooked' the results; indeed some of Mendel's own work is suspect for this reason).

Note that if we adopt the 5 (etc.) % level for our critical χ^2 we shall be wrong on 5 (etc.) % of occasions when the observed frequencies really do come from a population whose proportions in each class are as stated in our hypothesis.

11.3 Yates' correction

With this correction, instead of calculating χ^2 by (11.1) we use

$$\chi^2 \text{ (corrected)} = \frac{(|o_1 - e_1| - 0 \cdot 5)^2}{e_1} + \frac{(|o_2 - e_2| - 0 \cdot 5)^2}{e_2}$$

$$+ \dots + \frac{(|o_k - e_k| - 0 \cdot 5)^2}{e_k} \qquad (11.3)$$

and test against the (same) critical values of χ^2 with the ν of (11.2).

Note, however, that Yates' correction tends to over-correct (except in the binomial case of one set of n trials when $p = \frac{1}{2}$). Hence it is best used as follows: (i) If both χ^2 and χ^2 (corrected) are greater than χ^2 (critical), significant difference is established, (ii) if both are less, no significant difference has been established, (iii) if χ^2 is greater but χ^2 (corrected) less, take more observations.

Applying Yates' correction to the examples of Section 11.2 we have

$$\chi^2 \text{ (corrected)} = \frac{(|72-80| - 0 \cdot 5)^2}{80} + \frac{(|87-80| - 0 \cdot 5)^2}{80}$$

$$+ \frac{(|68-60| - 0 \cdot 5)^2}{60} + \frac{(|13-20| - 0 \cdot 5)^2}{20}$$

$$= \frac{(8-0 \cdot 5)^2}{80} + \frac{(7-0 \cdot 5)^2}{80} + \frac{(8-0 \cdot 5)^2}{60} + \frac{(7-0 \cdot 5)^2}{20} = 4 \cdot 28$$

11.4 Testing the proportion of successes in one set of n trials; confidence intervals

The χ^2 test plus Yates' correction enables us to test for significance using only the critical values of $\chi^2_{0 \cdot 05}$, $\chi^2_{0 \cdot 01}$, $\chi^2_{0 \cdot 001}$ for $\nu = 1$ (instead of extensive prepared binomial tables, though these are somewhat more accurate).

Thus, suppose s_0 successes are observed in n trials. We have only two classes, here, 'successes' (whose observed frequency $o_1 = s_0$) and 'non-successes' (with observed frequency $o_2 = n - s_0$). On the hypothesis that the expected proportion of successes should be p, the expected frequencies are $e_1 = np$, $e_2 = n - np = nq$. So,

$$\chi^2(\text{uncorrected}) = \frac{\{s_0 - np\}^2}{np} + \frac{\{n - s_0 - (n - np)\}^2}{nq}$$

$$= \frac{(s_0 - np)^2}{npq} \qquad (11.4)$$

$$\chi^2(\text{corrected}) = \frac{\{|s_0-np| - 0.5\}^2}{np} + \frac{\{|n-s_0| - (n-np)| - 0.5\}^2}{nq}$$

$$= \frac{\{|s_0-np| - 0.5\}^2}{npq} \qquad (11.5)$$

The number of classes is two, the number of relations between observed and expected frequencies is one (their sums are equal) and so $\nu = 2 - 1 = 1$. Table A.5 gives

$$\chi^2_{0.05}(\nu = 1) = 3.841, \quad \chi^2_{0.01}(\nu = 1) = 6.635,$$

$$\chi^2_{0.001}(\nu = 1) = 10.827 \qquad (11.6)$$

For example, suppose that 36 tosses of a coin give 12 heads and we wish to test the hypothesis that $p = \frac{1}{2}$. Here $n = 36$, $s_0 = 12$, and so

$$\chi^2 \text{ (uncorrected)} = \frac{(12-18)^2}{(36 \times \frac{1}{2} \times \frac{1}{2})} = 4.00$$

$$\chi^2 \text{ (corrected)} = \frac{(6-0.5)^2}{9} = 3.37$$

and the uncorrected test indicates significant differences at the 5% level but the corrected test does not (but since the correction is almost exact for $p = \frac{1}{2}$, we accept the latter verdict here, and do not reject the hypothesis that $p = \frac{1}{2}$).

Examples

11.4(i). In 180 throws of a die there are 40 sixes. Test the hypothesis that $p = \frac{1}{6}$.

11.4(ii). In 100 tosses there are 63 tails. Test the hypothesis that $p = \frac{1}{2}$.

Approximate 95% confidence limits for p can be obtained by finding the two values of p which satisfy

$$(|s_0-np| - 0.5)^2 = np(1 - p)\chi^2_{0.05}(\nu = 1) = 3.841np(1 - p) \qquad (11.7)$$

(start by solving the quadratic obtained by omitting the 0.5, which is formula (9.19) in another form, then finish solving by trial and error). The 99, and 99.9% limits obtained by using $\chi^2_{0.01}$ or $\chi^2_{0.001}$ are rather approximate.

11.4(iii). (a) When $s_0 = 8$, $n = 20$ show that the roots of (11.7) are 0.200, 0.636 (true values are 0.191 and 0.6395). (b) $s_0 = 3$, $n = 300$ show that the roots are 0.0026, 0.031 (true limits for p are 0.0036, 0.0292, see Example 9.8(iv) (e)).

11.5   The test for 'association' ('Two-way classified' or 'contingency tables')

Suppose it is thought that there is a tendency for fair-haired males to have blue eyes (in other words, that fair hair and blue eyes are 'associated'), and observations on 445 males give the results of Table 11.2(a). (Note that in this table males are classified in two ways, by hair- *and* eye-colour; further, since we have two classes of hair-colour, fair and not-fair, and two of eye-colour, blue and not-blue, Table 11.2(a) is called a 2 × 2 'contingency table'; with h

Table 11.2

(a) Observed frequencies				(b) Expected frequencies			
	Fair	Not-fair	Totals		Fair	Not-fair	Totals
Blue . .	75	103	178	Blue . .	58·4	119·6	178
Not-blue .	71	196	267	Not-blue .	87·6	179·4	267
Totals .	146	299	445	Totals .	146	299	445

classes of hair and k classes of eye colour the table would be called $h \times k$, see Table 11.3.)

We calculate the expected frequencies on the hypothesis that the same proportion of not-fair males as of fair males have blue eyes (called for short the 'null' or 'no-association' hypothesis). We do not know the exact proportion of blue-eyed males in the whole population, so we use the *observed* proportion, which is $\frac{178}{445}$. Further, the *observed* proportion of fair males is $\frac{146}{445}$. So our best estimate of the proportion of fair-haired blue-eyed males is (on the null hypothesis) $\frac{146}{445} \times \frac{178}{445}$. The expected *frequency* of fair-haired blue-eyed males is 445 times this proportion, i.e. $146 \times \frac{178}{445} = 58·4$. Similarly, the proportion (a) of not-fair blue-eyed males is $299 \times \frac{178}{445} = 199·6$,

Table 11.3

	21–30 yrs	31–40	41–50	51–60	61–70	Totals
Number 0	748	821	786	720	672	3747
of 1	74	60	50	66	50	300
Accidents 2 or more	40	35	29	21	22	147
Totals	862	916	865	807	744	4194

(b) of fair, not-blue-eyed is $146 \times \frac{287}{445} = 87.6$, and (c) of not-fair, not-blue-eyed is $299 \times \frac{287}{445} = 179.4$. So,

$$\chi^2(\text{uncorrected}) = \frac{(75-58.4)^2}{58.4} + \frac{(103-119.6)^2}{119.6} + \frac{(71-87.6)^2}{87.6}$$

$$+ \frac{(197-179.4)^2}{179.4} \tag{11.8}$$

$$= 4.72 + 2.30 + 3.15 + 1.54 = 11.71$$

$$\chi^2 \text{ (corrected)} = \frac{(16.6-0.5)^2}{58.4} + \frac{(16.6-0.5)^2}{119.6} + \frac{(16.6-05)^2}{87.6}$$

$$+ \frac{(16.6-0.5)^2}{179.4} \tag{11.9}$$

$$= 4.44 + 2.17 + 2.96 + 1.44 = 11.01$$

Now, there are four classes of males here; we have made the observed and expected totals of blue-eyed males equal, the totals of non-blue-eyed males equal, and the totals of fair males equal (*automatically*, then, the totals of not-fair males must be equal). So, really, there are three (independent) relations between the o_s and the e_s; hence $v = 4 - 3 = 1$ (this is true for all 2×2 contingency tables; the critical values of χ^2 for $v = 1$ are given in (11.6). Since both the corrected and uncorrected χ^2 of (11.8) and (11.9) are greater than $\chi^2_{0.001}(v = 1) = 10.83$, we reject the null (or no-association) hypothesis with very high confidence, and decide that association between fair hair and blue eyes is proved.

Examples

11.5(i). Experiments on anti-cholera inoculation gave these results. Can we be confident that inoculation is of value?

	Not-attacked	Attacked
Inoculated	276	3
Not-inoculated	473	66

11.5(ii). Test the no-association hypothesis for this data:

	Felt better	Did not
Took pill	64	14
Did not take pill	42	25

For the general $h \times k$ contingency table (i.e. h rows and k columns) the expected and observed totals in each row and column are equal and so

$$\nu = hk - (h + k - 1) = (h - 1)(k - 1) \qquad (11.10)$$

Otherwise the procedure is as above. Thus, suppose driver's ages and their accidents gave the results of the 3×4 Contingency Table 11.3. On the no-association hypothesis, the expected frequency of men in the 21–30 yrs., 0-accident class is $3,747 \times 862/4,194 = 770$, that in the 31–40 yrs., 1-accident class is $300 \times 916/4,194 = 65 \cdot 5$ and so on. Complete the test (N.B. $\nu = (3-1)(4-1) = 6$).

11.5(iii). Examiners A, B, C, D, and E graded papers as shown. Is there evidence of difference in grading?

	A	B	C	D	E
Distinction	17	25	36	29	41
Pass	146	107	104	93	87
Fail	32	19	24	31	20

11.6 The combination of χ^2 test results

In many problems observations have to be made on different groups (of people, say) and/or at different times (e.g. testing a treatment in several hospitals). The correct way of combining the results is simply to add the values of observed χ^2 and of ν for each experiment to get a *total* χ^2 and a *total* ν; the total χ^2 is then tested against critical χ^2 for the total ν.

Thus, the observations of Example 11.5(i) (for which $\chi^2 = 3 \cdot 27$, $\nu = 1$) were supplemented by 5 other sets (from workers on 5 other plantations), for which $\chi^2 = 9 \cdot 34$, $6 \cdot 08$, $2 \cdot 51$, $5 \cdot 61$, $1 \cdot 59$ with, of course, $\nu = 1$ in each case. Here, total $\chi^2 = 28 \cdot 40$ which is greater than $\chi^2_{0 \cdot 001}$ ($\nu = 6$) $= 20 \cdot 52$, see Table A.5. So the no-association hypothesis can be decisively rejected here (though the observed χ^2 for some individual tests is not significantly large).

Example

11.6(i). The values of observed χ^2, on the null hypothesis, for 4 different sets of 2×2 contingency table tests (N.B. $\nu = 1$ in each) were $2 \cdot 97$, $1 \cdot 95$, $3 \cdot 45$, $2 \cdot 48$. Test the individual and the total χ^2.

11.7 Goodness of fit for discrete distributions

(*a*) **Poisson.** The χ^2 test is often used to test whether observed data could come from a Poisson distribution; if so, then the data can be assumed to be random in origin (see Chapter 10). Table 11.4 gives the observations made by an apparatus recording the number of particles which arrived during 400 different periods each of a minute's duration.

Table 11.4

Number per minute s	0	1	2	3	4	5	6	7
Observed frequency o_s	102	143	98	42	9	5	1	0

The values of theoretical frequency e_s were calculated thus: The average value of s is

$$(0 \times 102 + 1 \times 143 + \ldots + 4 \times 9 + 5 \times 5 + 1 \times 6)/$$
$$(102 + 143 + \ldots + 5 + 1) = \tfrac{532}{400} = 1\cdot330.$$

Therefore, see (10.2), we put $e_s = 400\mu^s\, e^{-\mu}/s!$ with $\mu = 1\cdot330$ and calculate for $s = 0, 1, 2, 3$, to obtain $e_0, e_1, e_2, e_3 = 105\cdot8, 141\cdot1, 94\cdot0, 41\cdot8$. The frequencies for $s = 4, 5, 6$ or more are low so we group them into one class and, since the sum of the e_s has to equal the sum of the o_s, i.e. 400, the expected frequency in this class will be 17.3. So, finally, the observed and expected frequencies will be

o_s	102	143	98	42	15
e_s	105·8	141·1	94·0	41·8	17·3

Applying (11.1) we have

$$\chi^2 = (102\text{--}105\cdot8)^2/105\cdot8 + \ldots + (15\text{--}17\cdot3)^2/17\cdot3 = 0\cdot641$$

Now there are *two* relations between the o_s and e_s, namely their totals *and* their mean values are equal. There are five classes and so $\nu = 5 - 2 = 3$. Table A.5 gives $\chi^2_{0\cdot025}(\nu = 3) = 9\cdot35$ which is much larger than observed values of χ^2, $0\cdot641$, so we are quite confident that the theoretical Poisson distribution fits the observed data well.

Examples

11.7(i). Test the goodness of fit of the observed and frequency data of (a) Table 10.1, (b) the 3 sets of data of Example 10.2(ii).

11.7(ii). Test the hypothesis that (a) $\mu = 1\cdot30$, (b) $\mu = 1\cdot40$ for the data of Table 11.4.

11.7(iii). Fit a Poisson distribution to the following data on the number of particles seen in a small volume of a dusty gas in successive light flashes

Number of particles	0	1	2	3	4	5	>5
Observed frequency	34	46	38	19	4	2	0

Test the goodness of fit. (Data due to E. M. M. Badger, Gas, Light and Coke Company, quoted by O. L. Davies, *Statistical Methods in Research and Production*, Oliver and Boyd, London.)

(*b*) **Binomial (repeated sets of *n* trials).** Weldon, in a famous experiment, threw 12 dice repeatedly (26,306 times in all), and, defining a 'success' as either a 5 or a 6, counted *s*, the number of successes, at each throw of the 12 dice. The results are given in Table 11.5 together with two sets of expected frequencies whose nature will be described.

Table 11.5. DISTRIBUTION OF NUMBER OF SUCCESSES, *s* IN 26,306 THROWS IN 12 DICE

s	0	1	2	3	4	5	6	7	8	9	10 or more	Total
Observed frequency	185	1149	3265	5475	6114	5194	3067	1331	403	105	18	26,306
e_s	203	1217	3345	5576	6273	5018	2927	1254	392	87	14	26,306
e_s'	187	1146	3215	5465	6269	5115	3047	1330	424	96	16	26,306

On the hypothesis that the dice are unbiased the probability of either a 5 or a 6 is $\frac{1}{6} + \frac{1}{6} = \frac{1}{3}$. Hence the probability of *s* 'successes' among 12 thrown dice is, from (9.11), $^{12}C_s(\frac{1}{3})^s(\frac{2}{3})^{12-s}$, which when multiplied by 26,306 gives the row of expected frequencies e_s. For these,

$$\chi^2 = (185-203)^2/203 + (1,149-1,217)^2/1,217 + \ldots + (18-14)^2/14$$
$$= 35\cdot9 \qquad (11.11)$$

$$\nu = 11 \text{ (Number of classes)} - 1 \text{ (Observed and expected totals equal)}$$
$$= 10 \qquad (11.12)$$

Now, $\chi^2_{0\cdot001}(\nu = 10) = 23\cdot2$, which is much less than $35\cdot9$ and we decisively reject the $p = \frac{1}{3}$ hypothesis.

However, Weldon calculated the mean number of observed successes, which is

$$[185 \times 0 + 1,149 \times 1 + 3,265 \times 2 + 5,475 \times 3 + 6,114 \times 4 + \ldots]/26,306 = 4\cdot0524 \qquad (11.13)$$

The observed probability of success with a *single* die is therefore $4\cdot0524/12 = 0\cdot3377$. With $p = 0\cdot3377$ the expected frequency

$$e_s' = {}^{12}C_s(0\cdot3377)^s(0\cdot6623)^{12-s}$$

which gives the third row of the table. With these values

$$\chi^2 = (185-187)^2/187 + (1,149-1,146)^2/1,146 + \ldots + (18-16)^2/16$$
$$= 8 \cdot 20$$

We have two, relations between observed o_s and expected frequencies e_s namely their sums and their mean values are equal. So $\nu = 11 - 2 = 9$, here. Now Table A.5 gives $\chi^2_{0 \cdot 05}(\nu = 9) = 15 \cdot 5$, which is definitely greater than $8 \cdot 20$ and so we conclude that there is a good fit between observed and expected frequencies when $p = 0 \cdot 3377$. The physical reason why $p > \frac{1}{3}$ is that the 5 and 6 faces of a dice have slightly more scooped out of them than the 1 and 2 faces; this makes the 5 and 6 faces more likely to finish facing upwards.

Example

11.7(iv). Test the goodness of fit of the data of Table 9.2. Group the 4, 5, 6, 7, 8 or more frequencies together. (Note: Observed and expected totals and their means are equal.)

11.8 The χ^2 test applied to a continuous distribution

Suppose we wish to test whether a set of N observations $x_1, x_2 \ldots x_N$ could come from a normal population with specified mean μ and standard deviation σ. Now we know, see Table A.2, that $\frac{1}{10}$ of a normal population have variate-values between μ and $\mu + 0 \cdot 253\sigma$. In fact, any interval formed by two adjacent values of

$$-\infty, \quad \mu - 1 \cdot 282\sigma, \quad \mu - 0 \cdot 842\sigma, \quad \mu - 0 \cdot 524\sigma,$$
$$\mu - 0 \cdot 253\sigma, \quad \mu, \quad \mu + 0 \cdot 253\sigma, \quad \mu + 0 \cdot 524\sigma, \quad \mu + 0 \cdot 842\sigma,$$
$$\mu + 1 \cdot 282\sigma, \quad \infty$$

contains $\frac{1}{10}$ of the population. Suppose $o_1, o_2, \ldots o_{10}$ are the observed frequencies in these intervals (N.B. $o_1 + o_2 + \ldots + o_{10} = N$). The expected frequencies are $e_1 = e_2 = \ldots = e_{10} = N/10$. There are 10 classes and one relation ($\sum o_s = \sum e_s = N$); hence $\nu = 10 - 1 = 9$. We calculate $\chi^2 = \sum_{s=1}^{10} (o_s - e_s)^2/e_s$ and test against critical $\chi^2(\nu = 9)$ as before.

The intervals do not have to be 10 in number nor contain equal proportions of the population, but the principle remains the same.

If μ is determined from the observed mean $(x_1 + \ldots + x_N)/N$, then there are two relations between the o_s and e_s, and $\nu =$ Number of intervals (classes) $- 2$. If, further, the variance σ^2 is determined from s^2, the observed variance, then there is a quadratic relation (i.e. involving terms in $x_1{}^2$, $x_2{}^2$, etc.) between the o_s and e_s and the exact effect has so far proved too complicated to determine. However, if, using $v =$ Number of intervals—3, significance or non-significance is clearly established this verdict can be accepted, otherwise take more observations.

Exactly similar division into intervals can be used in the case of other continuous distributions.

12

Fitting lines and curves to data, least squares method

12.1 Nature of the problem

In scientific and technological work we often have to measure two quantities, one of which is subject to a certain amount of unpredictable variation, often called 'scatter' (e.g. the yield in some chemical reactions is frequently subject to scatter), whereas the other quantity can be determined beforehand exactly (e.g. the

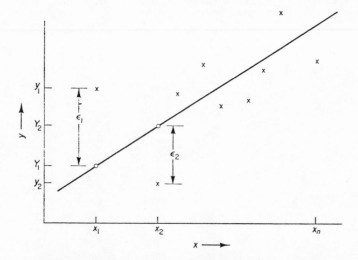

FIGURE 12.1. Diagram showing the errors $\varepsilon_1, \varepsilon_2, \ldots$ when a line is fitted to data with scatter

temperature or the duration of a reaction can be determined exactly, but the yield often cannot). The problem then is to find the true relation between the two quantities (e.g. exactly how does the yield vary with temperature).

If the data is plotted graphically (see Figs. 12.1 and 12.2) this problem becomes that of fitting a curve, or in some cases a straight line to the given set of plotted points. Such a curve, or line, is called by some statisticians a 'regression' curve or line, but others restrict this term, see Chapter 13. We now consider the simpler problem, that of fitting a straight line.

12.2 Fitting a straight line by 'least squares'

Given n pairs of observations (x_1, y_1), (x_2, y_2), . . . (x_n, y_n) where the y_i are subject to 'scatter' but the x_i are not, our problem is to choose values for a and b so that the line $y = a + bx$ fits these observations as closely as possible.

Now, if Y_1 is the y-value of that point on the line $y = a + bx$ whose x-value is x_1 (in other words $Y_1 = a + bx_1$), then we call $y_1 - Y_1$ the 'first error' and denote it by ε_1, see Fig. 12.1. Similarly, $y_2 - Y_2 = y_2 - (a + bx_2) = \varepsilon_2$, the 'second error', and so on up to the nth error $y_n - Y_n = y_n - (a + bx_n) = \varepsilon_n$. Then the best straight line to fit the data is that which makes the errors $\varepsilon_1, \varepsilon_2, \ldots$ ε_n as small as possible. Laplace, the famous French mathematician, put forward the suggestion that the best line is that line which makes the *sum* of the *squares* of the errors (call this sum S) as *small* as possible. Now if a and b are any values, the corresponding value of S is

$$S = \varepsilon_1{}^2 + \varepsilon_2{}^2 + \ldots + \varepsilon_n{}^2 = \{y_1 - (a + bx_1)\}^2 + \ldots$$
$$+ \{y_n - (a + bx_n)\}^2 \quad (12.1)$$

Laplace also solved the problem of finding the values of a and b for which S has the minimum or 'least' value. If $\bar{x}$ and $\bar{y}$ are the mean x and y-values, that is if

$$\bar{x} = (x_1 + x_2 + \ldots + x_n)/n, \quad \bar{y} = (y_1 + y_2 + \ldots + y_n)/n$$
$$(12.2)$$

then these values of a and b are given by the formulae

$$b' = \frac{\sum\limits_{i=1}^{n} (x_i - \bar{x})(y_i - \bar{y})}{\sum\limits_{i=1}^{n} (x_i - \bar{x})^2}, \quad a' = \bar{y} - b'\bar{x} \quad (12.3)$$

The equation for this 'best' line is often written

$$y - \bar{y} = b'(x - \bar{x}) \quad (12.4)$$

and it is usually called the 'least squares' line because the sum of the squares of its errors (i.e. its value of S) is the least of *all* the possible lines.

The following formulae simplify the computation of b':

$$\sum_{i=1}^{n} (x_i - \bar{x})(y_i - \bar{y}) = \sum_{i=1}^{n} x_i y_i - n\bar{x}\bar{y} \qquad (12.5)$$

$$\sum_{i=1}^{n} (x_i - \bar{x})^2 = \sum_{i=1}^{n} x_i^2 - n\bar{x}^2$$

Consider this example:

$$x_1, x_2, x_3, x_4 = 3, 4, 6, 7$$
$$y_1, y_2, y_3, y_4 = 12, 9, 10, 6$$

then

$$x_1 + x_2 + x_3 + x_4 = 20$$
$$y_1 + y_2 + y_3 + y_4 = 37$$
$$x_1^2 + x_2^2 + x_3^2 + x_4^2 = 110$$
$$x_1 y_1 + x_2 y_2 + x_3 y_3 + x_4 y_4 = 174$$

hence

$$\bar{x} = 5, \quad \bar{y} = 9{\cdot}25$$
$$b' = (174{-}185)/(110{-}100) = -1{\cdot}1$$
$$a' = 9{\cdot}25 - (-1{\cdot}1)5 = 14{\cdot}75$$

and so the least squares line in this case is

$$y = 14{\cdot}75 - 1{\cdot}1x$$

Examples

12.2(i) With the above data calculate ε_1, ε_2, ε_3, ε_4, S for each of the lines: (a) $y = 15 - 1{\cdot}3x$, (b) $y = 14 - 0{\cdot}9x$, (c) $y = 14{\cdot}75 - 1{\cdot}1x$ and show that case (c) gives the lowest value of S.

12.2(ii). Fit least squares lines to these two sets of data:

(a)

x	2	4	5	7	9	10
y	3	5	8	11	13	13

(b)

-3	-1	0	1	3	6
14	11	12	9	8	6

The proof of the above is as follows: By the differential calculus at a minimum $\partial S/\partial a = 0$ and $\partial S/\partial b = 0$. Differentiating S in (12.1) with respect to a and putting the result equal to zero, leads to the equation

$$na + (x_1 + x_2 + \ldots + x_n)b = y_1 + y_2 + \ldots + y_n \qquad (12.6)$$

and $\partial S/\partial b = 0$ leads to the equation

$$(x_1 + x_2 + \ldots + x_n)a + (x_1^2 + x_2^2 + \ldots + x_n^2)b$$
$$= x_1 y_1 + x_2 y_2 + \ldots + x_n y_n \quad (12.7)$$

The solution of these simultaneous equations for a and b leads to the formulae of (12.3).

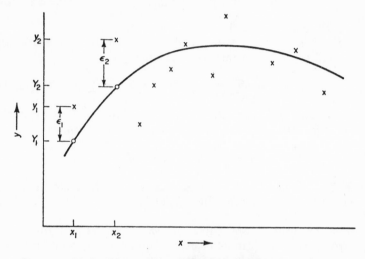

FIGURE 12.2. Diagram showing the errors ε_1, ε_2, . . . when a curve is fitted to data with scatter

12.3   Fitting quadratic or cubic curves by least squares

Data such as that of Fig. 12.2 is obviously better fitted by a curve than a line. A convenient type of curve to use is the 'quadratic' (also called the 'parabolic') curve which has the formula

$$y = a + bx + cx^2 \quad (12.8)$$

Again (see Fig. 12.2), if Y_1 is the y-value of the point on this curve whose x-value is x_1 (i.e. $Y_1 = a + bx_1 + cx_1^2$) then we call $y_1 - Y_1$ the first error ε_1. Similarly, $y_2 - Y_2 = y_2 - (a + bx_2 + cx_2^2) = \varepsilon_2$ the second error, and so on up to $y_n - Y_n = y_n - (a + bx_n + cx_n^2) = \varepsilon_n$. Again, we find the values of a, b, and c which make the sum of squares

$$S = \varepsilon_1^2 + \varepsilon_2^2 + \ldots + \varepsilon_n^2 \quad (12.9)$$

a minimum.

By using the differential calculus, it can be proved that these a, b, and c are then found by solving the simultaneous equations

(i) $$an + b\sum x_i + c\sum x_i^2 = \sum y_i$$

(ii) $$a\sum x_i + b\sum x_i^2 + c\sum x_i^3 = \sum x_i y_i \qquad (12.10)$$

(iii) $$a\sum x_i^2 + b\sum x_i^3 + c\sum x_i^4 = \sum x_i^2 y_i$$

where we sum from $i = 1$ to $i = n$ everywhere.

If $d\sum x_i^3$, $d\sum x_i^4$, $d\sum x_i^5$ are added to the left-hand sides of (i), (ii), (iii) in (12.10) then, together with the equation

$$a\sum x_i^3 + b\sum x_i^4 + c\sum x_i^5 + d\sum x_i^6 = \sum x_i^3 y_i \qquad (12.11)$$

their solution gives the values of a, b, c, d for the 'least squares' cubic curve fitting the data (a cubic curve has the formula

$$y = a + bx + cx^2 + dx^3)$$

It is best to have had experience of solving four simultaneous equations before trying this, though, luckily, it is rarely that a cubic curve has to be fitted.

Note that least squares curves do *not* in general go through the 'centroid' $(\bar{x}, \bar{y})$, and cannot be expressed in a neat general form like the least squares straight line (see (12.3)).

Example

12.3(i). Fit least squares quadratic and cubic curves to these sets of data:

(a) x	−4	−1	0	2	3
y	12	5	5	7	9

(b) x	1	3	4	5	6	8
y	−3	2	7	8	5	0

12.4   Fitting a general curve by least squares

We can fit a curve of the form

$$y = A_1 f_1(x) + A_2 f_2(x) + \ldots + A_k f_k(x) \qquad (12.12)$$

where $f_1(x), f_2(x), \ldots f_k(x)$ can be any k functions of x (e.g. we could have $f_1(x) = \log x$, $f_2(x) = \sin 7x$, $f_3(x) = x^3 + 5$, etc., though such bizarre combinations are unlikely). The A_i are constants and their values for the least squares curve, given the k functions $f_1(x), \ldots f_k(x)$, are found by solving the simultaneous equations

$$A_1\sum f_1(x_i) + A_2\sum f_2(x_i) + \ldots + A_k\sum f_k(x_i) = \sum y_i$$
$$A_1\sum x_i f_1(x_i) + \ldots + A_k\sum x_i f_k(x_i) = \sum x_i y_i$$
$$A_1\sum x_i^2 f_1(x_i) + \ldots + A_k\sum x_i^2 f_k(x_i) = \sum x_i^2 y_i$$
$$\vdots$$
$$A_1\sum x_i^{k-1} f_1(x_i) + \ldots + A_k\sum x_i^{k-1} f_k(x_i) = \sum x_i^{k-1} y_i \quad (12.13)$$

12.4(i). Fit a curve of the form: (a) $y = A_1 x + A_2 x^2$, (b) $y = A_1 x + A_2/x$ to this data

x	1	2	3	5	7	8
y	10	6	2	3	5	7

12.5 Assessing the fit by the coefficient of determination

Now, we call $\sum_{i=1}^{n} (y_i - \bar{y})^2$ the 'total variation' in y; $\bar{y}$ is, as usual, the mean y-value, i.e. $\bar{y} = (y_1 + \ldots + y_n)/n$. It can be proved that with least squares lines or least squares curves (of whatever type),

$$\sum(y_i - \bar{y})^2 = \sum(y_i - Y_i)^2 + \sum(Y_i - \bar{y})^2 \quad (12.14)$$

where Y_i is, as usual, the y-value of the point on the least squares line (or curve) whose x-value is x_i.

[For a non-least squares curve the relation is
$$\sum(y_i - \bar{y})^2 = \sum\{y_i - Y_i + (Y_i - \bar{y})\}^2$$
$$= \sum(y_i - Y_i)^2 + 2\sum(y_i - Y_i)(Y_i - \bar{y}) + \sum(Y_i - \bar{y})^2$$
and the middle term is not zero, in general, as it is for least squares lines or curves.]

We call $\sum(Y_i - \bar{y})^2$ the 'explained variation' because this is the variation of the points Y_i on the line (or curve), and we call $\sum(y_i - Y_i)^2$ the unexplained variation.

The ratio of the explained variation to the total variation, i.e. $\sum(Y_i - \bar{y})^2/\sum(y_i - \bar{y})^2$ is called the 'coefficient of determination' and it acts as a measure of the goodness of fit. Thus, the larger it is, the more of the total variation is accounted for by the explained variation. If this coefficient is 1, then the total variation is all accounted for by the explained variation.

Again, if, with a given set of data the coefficient of determination for the least squares line is low, but for the least squares quadratic curve is much higher, we consider that the quadratic curve fits the data better.

Examples

12.5(i). Show that with this data the coefficient of determination for the least squares line is <0.05, but for the least squares quadratic curve is >0.90:

x	1	3	4	5	7
y	−2	1	3	2	−4

However, if the two coefficients are not very dissimilar a more accurate test is needed and this we now discuss.

12.6   To test whether a line or curve fits the data

It is always possible that the variation in the observed y-values is entirely due to the 'scatter'. In other words, each y-value may be just the sum of a constant plus a random element (which we assume to be a random individual from some normal population). Call this the state of no relation between y and x.

We shall now give tests which determine whether this state of no relation is likely or whether there is evidence that some basic relation such as a straight line or quadratic curve exists between y and x (plus, of course, some random variation).

(*a*) **Test for a least squares line.** Here we first calculate the least squares a' and b', and then

$$s_1^2 = \sum_{i=1}^{n} (Y_i - \bar{y})^2, \quad s_2^2 = \left\{ \sum_{i=1}^{n} (y_i - Y_i)^2 \right\}/(n - 2) \quad (12.15)$$

where, of course, $Y_i = a' + b'x_i$. If $F = s_1^2/s_2^2 >$ critical $F(\nu_1 = 1, \nu_2 = n - 2)$ then we are confident that the least squares line fits the data better than the state of no relation between y and x. (Theory shows that if there really is no relation between y and x then s_1^2 and s_2^2 are two estimates of the *same* variance. This is improbable if $F = s_1^2/s_2^2 >$ critical F.) For computation it is easier to use the formulae

$$s_1^2 = b'^2(\sum x_i^2 - n\bar{x}^2), \quad s_2^2 = \{\sum y_i^2 - n\bar{y}^2 - s_1^2\}/(n - 2) \quad (12.16)$$

For example, if

$$x_1, x_2, \ldots x_5 = 0, 1, 3, 5, 6; \quad y_1, \ldots y_5 = 8, 5, 7, 2, 3$$

$\bar{x} = 3$, $\bar{y} = 5$, and

$$\sum x_i y_i - n\bar{x}\bar{y} = 54 - 75 = -21$$
$$\sum x_i^2 - n\bar{x}^2 = 71 - 45 = 26$$

hence $b' = -\frac{21}{26}$, $s_1^2 = \frac{21^2}{26} = 17{\cdot}0$

$$\sum y_i^2 - n\bar{y}^2 - s_1^2 = 151 - 125 - 17 = 9$$

and so $s_2^2 = \frac{9}{3} = 3{\cdot}0$; hence $F = s_1^2/s_2^2 = 17{\cdot}0/3{\cdot}0 = 5{\cdot}67$. But critical $F(\nu_1 = 1, \nu_2 = 3) = 10{\cdot}13$ at the 5% level and so we cannot be confident that this data can be represented by a straight line (however, 5 pairs of readings are rarely sufficient to give enough evidence, and further observations should be taken if possible).

Example

12.6(i). Apply this test to the data of Examples 12.2(ii) and 12.3(i).

(*b*) **Test for least squares quadratic and cubic curves.** If a, b, c for the least squares quadratic curve have been found (by solving (12.9)) then we calculate

$$S_1^2 = \sum(Y_i - \bar{y})^2/2, \quad S_2^2 = \sum(y_i - Y_i)^2/(n - 3) \quad (12.17)$$

where, of course, $Y_i = a + bx_i + cx_i^2$. If $F = S_1^2/S_2^2 >$ critical $F(\nu_1 = 2, \nu_2 = n - 3)$ we are confident that the quadratic curve represents the data better than the state of no relation.

For cubic curves, find the least squares a, b, c, and d, and then

$$T_1^2 = \sum(Y_i - \bar{y})^2/3, \quad T_2^2 = \sum(y_i - Y_i)^2/(n - 4)$$

where, of course,

$$Y_i = a + bx_i + cx_i^2 + dx_i^3$$

and if $F = T_1^2/T_2^2 >$ critical $F(\nu_1 = 3, \nu_2 = n - 4)$ we are confident that the cubic curve is better than no relation.

Example

12.6(ii). Apply the quadratic test to the data of Example 12.3(i).

(*c*) **Test for improvement of a quadratic curve over a line.** We sometimes wish to know whether the least squares quadratic curve fits the data better than the least squares line. We then calculate

$$F = (2S_1^2 - s_1^2)/S_2^2 \quad (12.18)$$

if this is greater than critical $F(\nu_1 = 1, \nu_2 = n - 3)$ we are confident that the quadratic curve fits better.

Again, if

$$F = (3T_1{}^2 - 2S_1{}^2)/T_2{}^2 > \text{critical } F(\nu_1 = 1, \nu_2 = n - 4)$$

we are confident that the least squares cubic curve is an improvement over the quadratic.

(d) Note on the above tests and on the fitting of curves. Unless it is *known* that the true relation between y and x is of a definite type, the simplest curve that fits the data reasonably well is usually accepted. It should be borne in mind that even when the true type of curve is known the estimated values of a, b, c, etc. (or of A_1, A_2, . . . in Section 12.4) may be very considerably in error unless n, the number of pairs of observations, is very large.

12.7 A simple alternative method of fitting a line

Consider first the case when n is even $= 2m$, say. Arrange x_1, x_2, . . . x_n in order, that is call the (algebraically) lowest value x_1, the next lowest value x_2, and so on. Group them into two halves, the lower x-values, i.e. x_1, x_2, . . . x_m and the upper x-values, i.e. x_{m+1}, x_{m+2}, . . . x_n. Find the 'centroids' $(\bar{x}_l, \bar{y}_l)$ of the lower group and $(x_u, \bar{y}_u)$ of the upper group. That is, calculate

$$\bar{x}_l = (x_1 + x_2 + \ldots + x_m)/m, \; \bar{y}_l = (y_1 + y_2 + \ldots + y_m)/m \tag{12.19}$$

$$\bar{x}_u = (x_{m+1} + x_{m+2} + \ldots + x_n)/m, \; \bar{y}_u = (y_{m+1} + y_{m+2} + \ldots + y_n)/n \tag{12.20}$$

Then joining the two centroids gives a simple alternative line to the least squares line whose fit does not usually seem worse, at least to casual inspection.

If n is odd $= 2p + 1$, say, the middle (median) x-value, x_{p+1}, presents a slight difficulty. It is usually put half in the lower group and half in the upper group. That is we calculate

$$\bar{x}_l = \frac{x_1 + \ldots + x_p + 0.5x_{p+1}}{p + 0.5}, \quad \bar{y}_l = \frac{y_1 + \ldots + y_p + 0.5y_{p+1}}{p + 0.5}$$

$$\bar{x}_u = \frac{0.5x_{p+1} + x_{p+2} + \ldots + x_n}{p + 0.5}$$

$$\bar{y}_u = \frac{0.5y_{p+1} + y_{p+2} + \ldots + y_n}{p + 0.5}$$

and, again, we join $(\bar{x}_l, \bar{y}_l)$ and $(\bar{x}_u, \bar{y}_u)$.

Example

12.7(i). Fit lines by this method to the data of Examples 12.2(ii) (*a*) and (*b*).

12.8 Least squares applied to three variables

Sometimes a quantity which is subject to scatter can depend on two other variables that can be fixed exactly (e.g. the yield of a chemical product can depend on the time of reaction *and* the temperature).

The least squares method can be used thus: Suppose we have a set of observations $(x_1, y_1, z_1), (x_2, y_2, z_2), \ldots (x_n, y_n, z_n)$ with scatter in z only, and suppose we wish to fit an equation

$$z = A + Bx + Cy \qquad (12.21)$$

to the data.

Again, the error $\varepsilon_1 = z_1 - (A + Bx_1 + Cy_1)$, and so on up to $\varepsilon_n = z_n - (A + Bx_n + Cy_n)$. The values of A, B, C, making $S = \varepsilon_1^2 + \varepsilon_2^2 + \ldots + \varepsilon_n^2$ a minimum, are given by solving

$$An + B \sum_{i=1}^{n} x_i + C \sum_{i=1}^{n} y_i = \sum_{i=1}^{n} z_i$$

$$A\sum x_i + B\sum x_i^2 + C\sum x_i y_i = \sum x_i z_i$$

$$A\sum y_i + B\sum y_i x_i + C\sum y_i^2 = \sum y_i z_i \qquad (12.22)$$

Example

12.8(i). Fit a curve of the form $z = A + Bx + Cy$, where there is scatter in z, only, to this data:

x	1	2	2	3	3	4	4	5
y	1	2	3	2	4	1	3	2
z	10	8	4	12	6	19	9	16

13

Regression curves and lines, correlation coefficient, normal bivariate distribution

13.1 The nature of regression curves and correlation

Suppose we wished to investigate the relation between the height of fathers and that of their eldest sons (when grown up). (Of course the reader will have noticed that tall fathers tend to have tall sons; but are the sons shorter or taller *on average* than their fathers? After all, some tall fathers have short sons and vice versa.) Suppose observations of a large number of heights of father and son have been made and that this data has to be analysed.

One way of dealing with it is to take all fathers in some small height range, say from 6 ft 1·0 in. to 6 ft 1·2 in., and find the mean height of their sons; and to repeat this for each small height range. We then plot the sons' mean height against the fathers' mean height, (for fathers in the height range 6 ft 1·0 in. to 6 ft 1·2 in. the mean sons' height for Englishmen is about 5 ft 10·6 in., the mean fathers' height is, of course, about 6 ft 1·1 in.). The curve passing through these points is called the 'regression curve for son's height on father's height' and Fig. 13.1 shows such a 'curve' (it turns out to be nearly a straight line).

Now, if we take sons in a given height range and find the mean height of their fathers and plot all such points we get a *different* curve called the 'regression curve for father's height on son's height'. This is usually different from the first regression curve, see Fig. 13.1. The reason for this difference can be seen by considering very tall fathers, say of height 6 ft 8 in. Their sons have a mean height about 6 ft 2 in. However, the majority of 6 ft 2 in. sons come from fathers nearer average height than themselves, simply because there are so *many* fathers near average height, and so *few* 6 ft 8 in. fathers.

Investigations of heights of father and son have been made by several investigators who all found that *both* regression curves were straight lines or nearly so. This is also true of very many observations of pairs of related measurements, especially in biology (the reason is discussed in Section 13.5).

Let the regression line for son's height (denoted by y) on father's height (denoted by x) have the equation

$$y - \bar{y} = B(x - \bar{x}) \tag{13.1}$$

where $\bar{x}$, $\bar{y}$ are the (overall) mean heights of fathers and sons. B is usually about 0·50, and for Englishmen $\bar{x} = \bar{y} = 5$ ft 8 in. Sir Francis Galton having obtained such a line, noted that it meant that fathers taller than the average by a certain amount had sons whose mean height was also above average but by only *half* this amount.

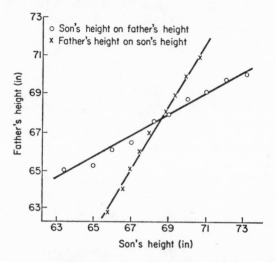

FIGURE 13.1. Regression lines for heights of father and son. (Data due to K. Pearson and A. Lee, *Biometrika*, 1903, reproduced by kind permission of the Biometrika Trustees)

Similarly, fathers below average by a certain amount have sons whose mean height is also below average but by half this amount. This tendency for sons to 'regress', i.e. go back towards the average (true for many other characteristics besides height) is the origin of the term 'regression' curve or line.

The regression line for father's height x on son's height y has the equation

$$y - \bar{y} = B'(x - \bar{x}) \tag{13.2}$$

where $B' = 1/B$ (the reason for this is given in Section 13.5).

Example

13.1(i). Find the mean son's height given a father's height of 6 ft 6 in., 6 ft 1 in., 5 ft 9 in., 5 ft 3 in., 4 ft 10 in.; find the mean father's height given a son's height of 6 ft 4 in., 6 ft 0 in., 5 ft 10 in., 5 ft 2 in., 4 ft 6 in.; take $\bar{x} = \bar{y} = 5$ ft 8 in., $B = 0·50$, $B' = 2·0$.

Regression curves can be found for many pairs of related observations, e.g. for height and foot-length of the same person, or for his age and blood pressure (this increases with age), or his age and highest audible pitch (this decreases with age), or for the yield per acre of potatoes and wheat in the counties of Britain, etc. These curves give more concise information about the relation between the two variables than the data itself. It is important to note that *both* the variables are subject to 'scatter' in the examples given above, *unlike* the curve fitting cases of Chapter 12 where only y was subject to scatter and x could be controlled precisely.

However useful curves are, it is desirable to have a quantitative measure of how closely the two variables are related and the 'correlation coefficient' now to be defined does provide such a measure (though it has limitations).

13.2   The correlation coefficient

Suppose we have N pairs of observations (x_1, y_1), (x_2, y_2), . . . (x_N, y_N). We first find the means $\bar{x}$, $\bar{y}$ of the x and y-values taken separately, i.e.

$$\bar{x} = (x_1 + x_2 + . . . + x_N)/N, \quad \bar{y} = (y_1 + y_2 + . . . + y_N)/N \tag{13.3}$$

then we calculate s_x^2, s_y^2 the observed x and y-variances, i.e.

$$s_x^2 = \{(x_1 - \bar{x})^2 + . . . + (x_N - \bar{x})^2\}/(N - 1),$$
$$s_y^2 = \{(y_1 - \bar{y})^2 + . . . + (y_N - \bar{y})^2\}/(N - 1) \tag{13.4}$$

Then we find what is called the 'covariance' c_{xy} of the x and y-values,

$$c_{xy} = \{(x_1 - \bar{x})(y_1 - \bar{y}) + (x_2 - \bar{x})(y_2 - \bar{y}) + . . .$$
$$+ (x_N - \bar{x})(y_N - \bar{y})\}/(N - 1) \tag{13.5}$$

The (observed) correlation coefficient r is defined thus

$$r = c_{xy}/(s_x s_y) \tag{13.6}$$

The correlation coefficient can be calculated directly from this formula

$$r = \frac{\sum_{i=1}^{N} (x_1 - \bar{x})(y_1 - \bar{y})}{\left[\sum_{i=1}^{N} (x_i - \bar{x})^2 \sum_{i=1}^{N} (y_i - \bar{y})^2\right]^{1/2}} \tag{13.7}$$

Note that in (13.6) s_x and s_y are always *positive* but c_{xy} may be positive or negative as will be shown below.

Examples

13.2(i). Calculate the covariance c_{xy} for the following sets of data
(a) (1, 10), (2, 11), (4, 9), (6, 7), (7, 3); (b) (2, 9), (4, 10), (5, 10),
(8, 12), (10, 13), (16, 18); (c) (−4, 9), (−2, 2), (0, 12), (3, 4).

13.2(ii). Find r given that s_x, s_y, c_{xy} = (a) 18, 8, 6; (b) 203, 89,
−117; (c) 0·142, 0·984, −0·043.

13.2(iii). Find the correlation coefficient r for each of the sets of
data of Example 13.2(i).

Just as the variance is a measure of the deviations from the mean
of a single set of observations such as x_1, x_2, . . . x_n, so the covariance is a measure of how the x and y-deviations vary *together*. Thus
if large positive x-deviations are associated with large negative
y-deviations and vice versa, c_{xy} will be large and negative, since there
will be many terms in the numerator of (13.5) which will be the
product of a large positive and a large negative term. Or, if the large
positive x and y-deviations and the large negative x and y-deviations
are associated together the covariance will be large and positive.
If there is no particular association, however, the terms in the
numerator of (13.5) will be of different signs and will tend to cancel
each other thus making the covariance small. Note that the correlation coefficient (see (13.6)) is really the *ratio* of the covariance c_{xy} to
the *geometric mean* of the x and y-variances, i.e. to $\sqrt{(s_x^2 s_y^2)}$.

It can be proved mathematically that the correlation coefficient r
can never be greater than $+1$ nor less than -1 (it cannot be $-1 \cdot 1$ for
example) *whatever* the x and y-values.

If we measure *every* pair of observations (e.g. the heights of *all*
fathers and eldest sons) we obtain the true or 'population' correlation
coefficient, which is usually denoted by ρ. The value of r obtained
from a sample is therefore only an estimate of ρ and so subject to
some error. We show how to test r for significance in Section 13.5
but to do this we first must discuss the normal bivariate distribution
in Section 13.3 in which ρ plays an essential role.

If the population correlation coefficient ρ has a value between
$+0 \cdot 8$ and $+1$ this indicates that there is a strong positive correlation
(i.e. the large positive x and y-deviations tend to be correlated or
associated together as are the large negative deviations). Values
between $-0 \cdot 8$ and -1 indicate strong 'negative correlation' (that
is large positive x-deviations are associated with large negative
y-deviations and vice versa). If ρ actually $= +1$ (or -1) we say
there is 'perfect' positive (or negative) correlation. Values around
$0 \cdot 5$ or $-0 \cdot 5$ indicate a fair amount of correlation; between $-0 \cdot 2$ and
$+0 \cdot 2$ indicate only weak correlation; and $\rho = 0$ means that there
is no correlation at all (with some reservations, see Section 13.6).

The following are some observed correlations: between heights or other characteristics (a) of parent and child, about 0·47, (b) between cousins about 0·26; between age and highest audible pitch (of the same person) about $-0·60$, a fairly strong negative correlation; between sunshine and rainfall in Hertfordshire, about $-0·2$, a weak negative correlation.

13.3 The normal bivariate distribution

This theoretical distribution is the natural extension of the normal population of Chapter 4 (where x is the variate) to the case of *two* variates x and y (hence the name 'bivariate'). It represents fairly well the distribution of many populations of related pairs of observations, especially many of biological origin (e.g. heights of father and son).

The basic formula of this distribution is as follows: The probability that, when a pair of observations are made at random, the x-value lies between X and $X + dX$ and the y-value between Y and $Y + dY$ is

$$\frac{1}{2\pi\sigma_x\sigma_y} \exp\left[-\tfrac{1}{2}\left\{\frac{(X-\mu_x)^2}{\sigma_x^2} - \frac{2\rho(X-\mu_x)(Y-\mu_y)}{\sigma_x\sigma_y} + \frac{(Y-\mu_y)^2}{\sigma_y^2}\right\}\right] dX\,dY \quad (13.7)$$

where ρ is the population correlation coefficient, μ_x, μ_y are the true means of the x and y-values and σ_x^2, σ_y^2 their true variances.

As in the normal distribution, X and Y may theoretically take any values between $-\infty$ and $+\infty$; however, the probabilities associated with large positive or negative values of X and/or Y are very small indeed. There are tables of the distribution available (e.g. see *Handbook of Statistical Tables* by D. B. Owen, Addison-Wesley, Reading, Massachusetts) but they are rarely needed.

The distribution has one very important property: If, for a fixed value of x the mean value of y is found, and this is done for every value of x, the resulting regression curve is a *straight line*; in fact it has the equation

$$y - \mu_y = (\rho\sigma_y/\sigma_x)(x - \mu_x) \quad (13.8)$$

We call this the regression curve for y on x. Similarly, the line

$$x - \mu_x = (\rho\sigma_x/\sigma_y)(y - \mu_y),$$

i.e.

$$y - \mu_y = (\sigma_y/\rho\sigma_x)(x - \mu_x) \quad (13.9)$$

is the regression curve for x on y (formed by the mean values of x for every fixed y). If observed regression curves are found to be straight lines or nearly so then this is an indication, though not a complete proof, that the underlying population is a normal bivariate one.

13.4 Estimating regression lines from samples

Given a sample of pairs of observations we estimate the constants μ_x, μ_y, σ_x, σ_y, and ρ by the corresponding observed quantities $\bar{x}$, $\bar{y}$, s_x, s_y, and r as calculated in Section 13.2. Substituting these in (13.8) we obtain the estimated regression line for y on x, namely

$$y - \bar{y} = (rs_y/s_x)(x - \bar{x}) \qquad (13.10)$$

This is the *same* line as the *least squares* line (12.4) (which is calculated on the assumption that there is scatter in y only), and this can be proved thus

$$\frac{rs_y}{s_x} = \frac{c_{xy}}{s_x^2} = \frac{\sum(x_i - \bar{x})(y_i - \bar{y})}{\sum(x_i - \bar{x})^2} = b' \qquad (13.11)$$

The estimated regression line for x on y is

$$x - \bar{x} = (rs_x/s_y)(y - \bar{y})$$

i.e.   (13.12)

$$y - \bar{y} = (s_y/rs_x)(x - \bar{x})$$

which is the same as we should find if we found the least squares line assuming scatter in x only. Note that we sometimes write $rs_x/s_y = \beta_1$ and $rs_y/s_x = \beta_2$ and call β_1 and β_2 the 'regression coefficients'. Note, too, that $\beta_1\beta_2 = r^2$.

If $s_x = s_y$ the slope of the regression line (13.10) equals the reciprocal of the slope of (13.12), i.e. $s_y/(rs_x) = 1/(rs_y/s_x)$. This explains why $B' = 1/B$ in Section 13.1 (the variance of father's and of son's heights are practically equal). Note too that if $r = +1$ or -1 the slopes of both regression lines (13.10) and (13.12) are the same namely s_x/s_y or, if $r = -1$, $-s_x/s_y$. This means that for perfect correlation the regression lines *coincide*.

Example

13.4(i). Give the estimated regression lines for each of the sets of data of 13.2(i).

13.5 Significance tests for r

On the assumption that each pair of observations (x, y) is selected at random from a normal bivariate distribution, we can give tests to decide whether r is so large ($+$ve or $-$ve) that the hypothesis that $\rho = 0$ (i.e. there is *no* correlation between x and y) can be rejected. The test function is

$$c = |r|\sqrt{(N-1)} \tag{13.13}$$

If $c > 1.96, 2.58$, or 3.29 we can reject the hypothesis at, about, the 5, 1, or 0.1% probability level ('about' because $r\sqrt{(N-1)}$ is distributed normally to an approximation which is only very good if N is large).

The above gives us a method of deciding that very probably there is some association or correlation between x and y. We can test a hypothesis about the strength of this association thus: given the hypothesis that the true correlation coefficient has a specified value ρ', we first calculate

$$z = \tfrac{1}{2}\log_e\left\{\frac{1+r}{1-r}\right\}, \quad z_0 = \tfrac{1}{2}\log_e\left\{\frac{1+\rho'}{1-\rho'}\right\} \tag{13.14}$$

and then the test function

$$c = |z - z_0|\sqrt{(N-3)} \tag{13.15}$$

and if $c > 1.96, 2.58, 3.29$ we reject the hypothesis at, about, the 5, 1, 0.1% level.

Examples

13.5(i). Test the hypothesis that $\rho = 0$ given (a) $r = 0.43$, $N = 10$; (b) $r = -0.37$, $N = 50$; (c) $r = -0.021$, $N = 290$.

13.5(ii). Test the hypothesis that (a) $\rho = 0.30$ when $r = 0.53$, $N = 145$; (b) $\rho = -0.40$, when $r = -0.28$, $N = 50$.

Professor E. S. Pearson has shown that these tests are reasonably valid even if the true distribution of x and y differs somewhat from the normal bivariate.

13.6 Some notes on the correlation coefficient

(a) **Relation to the coefficient of determination.** The following can be proved mathematically: the 'coefficient of determination' (i.e. explained variation/total variation, see Section 12.5) of the estimated regression line for y on x (which, see Section 13.4, is also the least squares lines assuming scatter in y only) *equals* r^2. The coefficient of determination of the estimated regression line for x on y *also equals* r^2.

(b) **Linear and non-linear correlation coefficients.** The correlation coefficient r measures well the correlation between x and y if the true regression curves are *lines*; and, for this reason, is often called the 'linear' (also, the 'product-moment') correlation coefficient. If one or both of the regression curves are quadratic (or more elaborate still), r can be nearly zero even though x and y are closely related. There is no general measure which can be used in all such cases, but if a quadratic curve fits the data reasonably, the square-root of the coefficient of determination can be used as a non-linear correlation coefficient, but little research has been carried out on this topic.

(c) **Correlation and causal connection.** A significant value of r does not necessarily imply that there is a *causal* connection between x and y. Thus, there is a positive correlation between expectation of life and number of registry office marriages (both have increased steadily since 1900), but no one supposes that a registry marriage will lengthen his life. Again, some opponents of the Doll Report conclusion that smoking causes lung cancer have argued thus: the high correlation between heaviness of smoking and incidence of lung cancer may be due to a hereditary factor which makes certain people both like smoking and be disposed to lung cancer.

If a connection between two quantities is suspected, a number of pairs of observations are made and r calculated. If small the idea is dropped or considered of little practical account. If r is large, then the next step is to investigate the causal mechanism, by scientific experiment, etc.

(d) **Correlation between three variables.** The notion of correlation can be extended to three quantities (e.g. height, weight, and foot-length of the same person) which may be related together. The relevant formulae are somewhat lengthy and suitable text-books should be consulted.

13.7 Spearman's coefficient of rank correlation

Sometimes we can only *rank* objects (or persons) in order of preference, e.g. beauty queens, textile cloths (which are often ranked on a mixture of appearance and 'feel'). Spearman devised a co-efficient (there are others) which measures the correlation between the rankings made by two persons. Suppose, for example, two experts A and B have ranked seven different textile cloths thus:

Table 13.1

Cloth number	1	2	3	4	5	6	7
A's ranking	6	4	1	3	7	5	2
B's ranking	3	4	2	5	7	6	1
Difference in ranking	+3	0	−1	−2	0	−1	+1

The Spearman Rank correlation coefficient ρ_s is defined thus:

$$\rho_s = 1 - 6\{\sum(\text{Difference})^2\}/(n^3 - n) \qquad (13.16)$$

where n is the number of objects ranked. For Table 13.1 we have $n = 7$, and

$$\rho_s = 1 - 6\{3^2 + 0 + (-1)^2 + (-2)^2 + 0 + (-1)^2 + 1^2\}/(7^3 - 7)$$
$$= 5/7$$

An approximate test of significance is this: calculate

$$t_s = \rho_s\{(n - 2)/(1 - \rho_s^2)\}^{1/2} \qquad (13.17)$$

and if $t_s >$ critical t $(\nu = n - 2)$ for the *one-tailed* test, we are confident that there is some agreement or 'concordance' between the rankings. For Table 13.1 we have

$$t = (5/7)\{5/(1-5^2/7^2)\}^{1/2} = 2\cdot48$$

which is greater than $2\cdot015$, the one-tailed critical t for $\nu = 5$ at the 5% level, see Table A.3. Hence we conclude that there is some concordance between the experts.

The basis of this test is this: If the rankings were made at random some differences would be large and this would make ρ_s appreciably less than 1, and it might even be negative. But, if there is concordance the differences will be small and ρ_s will be close to 1. So, if ρ_s is sufficiently close to 1 we are confident that there is some concordance. Perfect concordance would make all differences zero and, then, $\rho_s = +1$. Perfect 'discordance', i.e. when A ranks last what B ranks first, etc., makes $\rho_s = -1$. Finally, the test is one-tailed because significance is only indicated if ρ_s is positive and close to $+1$, if it is negative the concordance is more likely to be a discordance.

This test can be applied to see whether there is any correlation between pairs of *quantitative* observations (it is an alternative to the linear correlation coefficient). Thus, given the data $(0, 0)$, $(22, 1)$, $(23, 3)$, $(48, 4)$, $(25, 16)$, $(26, 18)$ we first rank the x and y-values separately and then replace each x or y-value by its rank. We give the lowest x or y-value rank 1, the next lowest rank 2 and so on, and, thus, the above data becomes $(1, 1)$, $(2, 2)$, $(3, 3)$, $(6, 4)$, $(4, 5)$, $(5, 6)$. Hence

$$\rho_s = 1 - 6\{0 + 0 + 0 + 4 + 1 + 1\}/(6^3 - 6) = 29/35 = 0\cdot829$$

and

$$t_s = 0\cdot829[4/\{1 - (0\cdot829)^2\}]^{1/2} = 2\cdot97$$

which is greater than $2\cdot776$, the two-tailed critical t for $\nu = n - 2 = 6 - 2 = 4$ at the 5% level. (N.B. Two-tailed values are used if a negative correlation is to be considered significant as well as a positive correlation.) The test shows that there is an appreciable positive correlation between the variables.

The linear correlation coefficient r is $+0.278$ for this data, and the test function c of Section 13.5 is 0.622, which is *much less* than 1.96. The reason for the failure of the test of Section 13.5 to detect significant correlation here, is that this test depends on the data coming from something like a normal bivariate distribution. The ranking method is, however, virtually *independent* of the distribution. The next chapter describes a variety of tests based, in essence, on the idea of using ranking.

Examples

13.7(i). The 8 finalists in a beauty competition are ranked by two judges as shown; find ρ_s and test for significance:

A	5	8	1	2	4	6	7	3
B	4	7	2	5	3	8	6	1

13.7(ii). By the ranking method calculate Spearman's coefficient and test it for significance for each set of data in Example 13.2(i).

14

Some distribution-independent (or 'distribution-free' or 'non-parametric') tests

14.1 Nature of the tests

The significance tests of Chapter 6 are applicable only if the observations come from distributions not far from normal. We now give tests which, in general, are independent of the type of distribution, and for this reason are sometimes called 'distribution-free' or 'non-parametric' tests. They are based on arranging the observations in order or on ranking them in some way.

We now give some of these tests which determine significant difference between two samples.

14.2 The 'run test' for the difference between two samples (a location and dispersion sensitive test)

Let $x_1, x_2, \ldots x_m$ be an 'ordered' sample (i.e. $x_1 \leqslant x_2 \leqslant \ldots \leqslant x_m$) and $y_1, y_2, \ldots y_n$ be a second sample also 'ordered'. For example, we might have $x_1, \ldots x_4 = 3, 5, 7, 10$; $y_1, y_2, y_3 = -2, 0, 6$. Then we write out the whole set of observations in order, thus

$$-2 \quad 0 \quad 3 \quad 5 \quad 6 \quad 7 \quad 10 \qquad (14.1)$$
$$y_1 \quad y_2 \quad x_1 \quad x_2 \quad y_3 \quad x_3 \quad x_4$$

Defining a 'run' as a sequence of one or more x's or one or more y's, we count the number R of runs. In (14.1) there are four runs, namely, $(y_1 y_2)$, $(x_1 x_2)$, (y_3), $(x_3 x_4)$, so $R = 4$.

If $m \geqslant 11$, $n \geqslant 11$ we then calculate

$$r = \frac{2mn}{m + n} + 1, \quad \sigma_r^2 = \frac{2mn(2mn - m - n)}{(m + n)^2(m + n + 1)} \qquad (14.2)$$

and, if $(\mu_r - R)\sigma_r \geqslant 1.645$, 1.96, 2.33, or 2.58 we can reject the hypothesis that the samples come from the same population at the 5, 2·5, 1·0, or 0·5% level. When $m \leqslant 10$ or $n \leqslant 10$, if R is less than the critical value as given in tables (e.g. D. B. Owen) the hypothesis is rejected.

The basis of the test is this: If the samples really come from the same population the x's and the y's will be well mixed and R will be appreciable. If, however, the samples come from two populations well separated in 'location' so that all the y's are greater than all the

x's then there may be only a run of x's and a run of y's so that $R = 2$. Again, if the y population is widely spread out (or 'dispersed') and the x population concentrated there may be only a run of y's, a run of x's and another run of y's, so that $R = 3$. Hence, a *low* value of R indicates that the populations differ in location and/or dispersion.

Example

14.2(i). Test the following pairs of samples for difference: (a) 0, 4, 5, 7, 8, 9, 10, 11, 14, 15, 20, 21; 12, 13, 16, 17, 19, 22, 24, 25, 27, 30, 34, 37; (b) -45, -40, -31, -17, -9, 4, 11, 22, 23, 25, 31, 41, 50; -11, -10, -2, 0, 7, 8, 9, 10, 12, 13, 18.

14.3 The median test (sensitive to location) for two samples

The median of a distribution is that variate-value which divides the distribution halfway, i.e. half the distribution (population) have lower and half have higher variate-values. If the distribution is *symmetrical* the median and the mean are the same, but for *skew* distributions this is not so (e.g. consider the simple population consisting of three individuals with variate-values 0, 1, 20; its median is 1 but its mean is $(0 + 1 + 20)/3 = 7$).

The test now described determines primarily whether the medians of the populations from which the samples come are well separated or not. It is little affected by greater dispersion or spread in one population than the other.

We first write out the observations in both samples in order and find the median of this set. Thus, in (14.1), 5 is the median (there are 3 lower and 3 higher observations). If the total number of observations is even the median is taken to be halfway between the two middle observations. If this total number is odd the median observation is removed and the case reduced to the even case. Then the number m' of x's and the number n' of y's greater than the median is noted. If the test function

$$M = (|2m' - m| - 1)^2/m + (|2n' - n| - 1)^2/n \qquad (14.3)$$

is *greater* than $\chi^2_{0.05}(\nu = 1)$, see Table A.5, we reject, at the 5% level, the hypothesis that the samples have the same median.

Thus, suppose we have two samples A and B consisting of the observations $A = -9$, -7, -6, -3, 0, 1, 9; $B = 3$, 6, 7, 10, 11, 14, 15, 16, 20, 23. When A and B are written out in order as a single set we see that 7 is the median observation. Removing 7 then 6 and 9 are the middle observations of the single set and the new median is therefore $\frac{1}{2}(6 + 9) = 7.5$. Since 9 is the only observation of A greater than 7.5 we have $m' = 1$. Similarly $n' = 7$, while $m = 7$

and $n = 9$ (since one observation has been removed from B). Hence

$$M = (|2 - 7| - 1)^2/7 + (|14 - 9| - 1)^2/9 = 16/7 + 16/9 = 4·16$$

which is greater than $3·84 = \chi^2_{0·05}(\nu = 1)$. So we *reject* at the 5% level the hypothesis that median A = median B.

The basis of the test is this: if the populations have the same median there would be a probability of $\frac{1}{2}$ that each x and each y is greater than this median and m' and n' would be close to $\frac{1}{2}m$ and $\frac{1}{2}n$.

If, however, the populations are well separated most of the x's will be either well above or well below the median of the combined set of observations. Hence m' will be large and n' small or vice versa; in either case M will be appreciable. The test gives no information if m and n are below 6; preferably they should be 12 or more.

Examples

14.3(i). Apply this test to the data of Example 14.2(i).

We can subtract a constant k from the sample with more observations above the median and then we can test the new pair of samples. The two values of k for which the test function M = critical χ^2 are the confidence limits for the difference between the population medians (N.B. $\chi^2_{0·05}$ gives the 95, $\chi^2_{0·01}$ the 99% confidence limits).

Taking the two samples given above, if we put $k = 6·1$ and subtract from sample B we get $-3·1, -0·1, +0·9, 3·9, \ldots = B_1$, say. Writing A and B_1 in order as a single set, we see that 1 is the median observation. Removing it, we have $m' = 1, m = 6, n' = 7, n = 10$. Hence

$$M = 9/6 + 9/10 = 2·4 < 3·84 = \chi^2_{0·05}(\nu = 1)$$

so we are not confident that the medians of A and B_1 are different. However, subtracting $k = 5·9$ from B we have $-2·9, +0·1, 1·1, 4·1, \ldots = B_2$ say. Writing A and B_2 as a single set, then $1·1$ is the median. Removing it, we have $m' = 1, m = 7, n' = 7, n = 9$. Hence

$$M = 16/7 + 16/9 = 4·16 > 3·84$$

and we are confident that the medians of A and B_2 *are* different. In fact if k is just greater than 6 we shall get no significant difference, but if k is just less than 6 we shall get a difference. So we can say we are confident that median B − median $A \geqslant 6$ at the 5% level. A similar argument with k just greater than or just less than 23, shows that, at the 5% level, median B − median $A \leqslant 23$.

14.3(ii). Subtract (*a*) 10, (*b*) 25, (*c*) 31 from each observation in the second sample of Example 14.2(i) (*b*) and test the new pair.

14.3(iii). Subtract 1·9, 2·1, 22·9, 23·1 from the second sample of Example 14.2(i) (*a*) and test the new pair in each case.

14.3(iv). Find 99% confidence limits for the difference between the population medians for the samples of Example 14.2(i) (*a*).

14.4 The sign test (paired comparisons)

To determine which of two varieties of the same plant gives the higher yield, we can grow them in pairs in a number of localities. Suppose the yields are $(x_1, y_1), \ldots (x_N, y_N)$. To test the 'null' hypothesis that there is no difference between the two yields we (i) find $x_1 - y_1, x_2 - y_2, \ldots x_N - y_N$, (ii) count the numbers p and m of plus and minus signs among these differences (we omit ties, i.e. cases where $x_i = y_i$), (iii) we compute

$$S = (|p - m| - 1)^2/(p + m) \qquad (14.4)$$

If $S > \chi^2_{0.05}$ ($\nu = 1$) we reject the null hypothesis at the 5% level, (if $S > \chi^2_{0.01}$ we reject at the 1% level).

Example

14.4(i). Apply this test to the following pairs of observations: (24, 15), (38, 24), (22, 28), (32, 31), (29, 28), (33, 29), (29, 31), (28, 28), (27, 25), (26, 26), (27, 31), (25, 33).

The basis of the test is this. If x and y are equally good, then

$$\Pr(x > y) = \tfrac{1}{2} = \Pr(y > x).$$

If there are more plus signs than would be expected on this probability of $\tfrac{1}{2}$, then we can reject the null hypothesis.

Again, as in Section 14.3, we can subtract any number k from each of the observations in one sample and test the new pair of samples. The two values of k for which $S = $ critical χ^2 give confidence limits for the difference between the median of the two yields.

Examples

14.4(ii). Subtract (*a*) 2, (*b*) 4 from the first observation of each pair in Example 14.4(i) and test the new signs formed.

14.4(iii). Find confidence limits for k for the data of Example 14.4(i).

The sign test can be used to test the hypothesis that the median of the population, from which we have drawn a *single* sample only, could equal k. If $x_1, \ldots x_N$ are the observations, we find p, m the number of plus and of minus signs amongst $(x_1 - k), (x_2 - k), \ldots (x_N - k)$ with zeros omitted. If

$$S = (|p - m| - 1)^2/(p + m) \geqslant \chi^2_{0.05}(\nu = 1)$$

we can reject this hypothesis. Confidence limits for k can be found by finding k_1 and k_2 such that $S = \chi^2_{0.05}$ in each case.

14.4(iv). Find confidence limits for the median given the sample 0, 8, 12, 13, 15, 17, 20, 21, 26, 30.

14.5 Wilcoxon's signed rank test

This improves upon the sign test by giving the larger differences greater 'weight'. To do this we: (i) rank the differences *without regard* to sign, i.e. smallest magnitude difference is given rank 1, next smallest rank 2, and so on; equal differences are given half each of the combined rank, etc., (ii) replace the original signs, (iii) find the total sum T of the positive *or* negative ranks, whichever is the *smaller*. If T is *less* than the critical value (see Table 14.1 or equation

Table 14.1. CRITICAL VALUES OF T AT THE 5% AND 1% LEVELS

n	6	7	8	9	10	11	12	13	14	15	16	17	18	19	20	21	22	23	24	25
5%	0	2	4	6	8	11	14	17	21	25	30	35	40	46	52	59	66	73	81	89
1%	–	–	0	2	3	5	7	10	13	16	20	23	28	32	38	43	49	55	61	68

(Reproduced by permission of F. Wilcoxon, and the American Cyanamid Co., Stamford, Conn.)

(14.5)) we reject the hypothesis that there is no difference between the x and y values.

Suppose we have the following pairs of observations: (16, 8), (18, 9), (12, 13), (18, 11), (21, 10), (15, 13), (12, 16), (16, 10), (20, 11). The differences in order of magnitude with their ranks and signs are

	$12 - 13 = -1$	$15 - 13 = +2$	$12 - 16 = -4$
Rank	1	2	3
Sign	—	+	—

	$16 - 10 = +6$	$18 - 11 = +7$	$16 - 8 = +8$
Rank	4	5	6
Sign	+	+	+

	$18 - 9 = +9$	$20 - 11 = +9$	$21 - 10 = +11$
Rank	7·5	7·5	9
Sign	+	+	+

The sum of the negatively signed ranks is

$$1 + 3 = 4 < \text{crit } T(n = 9) = 6$$

at the 5% level (see Table 14.1); so at this level (but not at the 1% level) we reject the hypothesis that there is no significant difference between the pairs of readings.

For $n > 25$ we compute

$$\frac{\frac{1}{4}n(n + 1) - T}{[\frac{1}{12}n(n + \frac{1}{2})(n + 1)]^{\frac{1}{2}}} \tag{14.5}$$

and, if this is greater than 1·96, 2·58, or 3·29 we reject the 'null' hypothesis at the 5, 1 or 0·1 % level.

Again, k may be subtracted from one set of observations and the new pair tested.

Example

14.5(i). Apply Wilcoxon's test to the data of Example 14.4(i), (*a*) without alteration, (*b*) after subtracting 5 from the first observation in each pair.

14.6   Some notes on distribution-independent tests

These tests are fairly simple, but are less efficient than those based on a *known* distribution, e.g. the normal (though for large n at least, Wilcoxon's test is only slightly inferior when the true distribution *is* normal).

These methods have been extended to the analysis of other problems, e.g. two-way classified data (defined in Chapter 7).

15

Note on sampling techniques and quality control

15.1 Random sampling of a finite number of objects

If we have a finite number of articles or plants and wish to select some at random, then care must be taken. For example, selecting by eye has been proved unsatisfactory, particularly for plants. Even the most careful experimenter seems to have a bias in some direction, and, unconsciously, tends to select those which are larger, or smaller, or greener, etc.

However, a very simple and reliable technique has been developed. The articles are numbered 1, 2, 3, . . . in *any* order. Then a table of prepared 'random numbers' is opened, and starting at an arbitrary point, the numbers are read off and are used to select the articles as follows:

The first row of the random numbers table (*Handbook of Statistical Tables*, by D. B. Owen, Addison-Wesley, Reading, Mass.), is

1,368 9,621 9,151 2,066 1,208 2,664 9,822 6,599 5,911 5,112

Suppose we have to select at random 10 plants out of a set of 77. We number the plants 1 to 77 in any manner. Then we select consecutive pairs of digits from the random numbers table starting at the arbitrary point 2,066. Thus we get

20, 66, 12, 08 = 8, 26, 64, [98], 22, 65, [99], 59, 11

The numbers greater than 77 are omitted, and so the ten plants with numbers 20, 66, 12, 8, 26, 64, 22, 65, 59, 11 are those actually observed.

If there were 863 plants, we would take sets of *three* digits and thus, starting at 1,368, plants numbered

136, [896], 219, 151, 206, 612, 082 = 82, 664, etc.,

would be chosen (896 is omitted of course).

15.2 Sampling continuous populations

With continuous populations, e.g. liquids in bulk, solids, etc., the problem of sampling becomes more complicated. Though, if the liquid can be stirred thoroughly then any small volume taken from it will have exactly the same composition as the whole, and the problem is then very easy. If stirring is not possible, as with solids, the bulk can be divided theoretically into small units of roughly equal volume, the units can be numbered and the random number technique of Section 15.1 used.

Unfortunately, this fails when units inside the bulk are not accessible (e.g. the cotton in a bale cannot be observed until the bale is broken up, but a decision as to its quality has, nevertheless, to be taken). The only resource, in such cases, is to conduct occasional experiments to find what 'bias' if any, results from observing accessible units only.

15.3   Stratified (Representative) sampling

In some cases we know beforehand that there are, or may be, differences between different parts or 'strata' of the population; and so, we include observations from each stratum in proportion to its size. Thus a public opinion poll might arrange to interview 100 people from Yorkshire and only 60 from Surrey, if the populations of these counties are in that ratio. (The random number technique applied to the whole population might result in, say, only 73 York-shiremen, but 81 Surrey men being interviewed.) Inside each county there might be a further stratification, say according to income, and in Yorkshire, 10 wealthy, 30 not so wealthy, 50 about average, and 10 poor people might be interviewed (actually, the determination of a person's income is not easy, and the Public Opinion Polls have some difficulty with it).

Similarly, where we are sampling from a consignment consisting of articles from several different sources, we often arrange to examine the same *proportion* of articles from each source (this is sometimes referred to as 'representative sampling').

Though stratification presents some complications, occasionally it simplifies the procedure. Thus, with many plants in many rows, rather than number every plant, it is easier to number the rows and select some by the random number technique of Section 15.1; then to number 'sections' in each of the selected rows, and again select some of these by the random number technique; and, finally to number the plants in each selected section and apply the random number technique to them.

15.4   Quality control

Where a product (e.g. cotton thread) is being manufactured, in order to keep the product of a standard quality, it is desirable that some property or quality (e.g. breaking strength) of the product should be measured and used as an indication that all is well or not. In manufacturing most products there are a number of sources of small variation, and so, the property measured is usually distributed *normally* about its mean value. The properties of the normal distribution can therefore be used to provide effective quality control.

This is usually done along these lines: Suppose the property measured is known from previous experience to have a mean μ and a standard deviation σ. Then a sample of n specimens of the product (n may of course be small, even 1) are selected at random (by the techniques of Section 15.1 or Section 15.2) and the mean of the observed measurements is plotted on a graph or chart; this usually has heavy lines at ordinate values of $\mu + 1.96\sigma/\sqrt{n}$ and $\mu - 1.96\sigma/\sqrt{n}$. These are called the 'warning lines' and other heavy lines at $\mu + 3.29\sigma/\sqrt{n}$ and $\mu - 3.29\sigma/\sqrt{n}$, are called the 'action lines' (the names vary from place to place but the meanings are the same). If a sample mean falls outside the 'warning lines', it is as well to look at the immediately previous sample means; because, by the properties of normal populations only 1 in 40 sample means should be greater than $\mu + 1.96\sigma/\sqrt{n}$ (or, less than $\mu - 1.96\sigma/\sqrt{n}$). If a mean falls outside the 'action lines' then probably something is wrong (because only 1 in 100 means should do this).

Of course, this regular recording has the advantage of presenting a ready picture of recent production, but without the warning and action lines it would not be possible to sort out real drifts from the scatter which is present.

Besides the sample mean (which measures the average quality) the sample range (Greatest − Smallest observation) can be found. This gives a measure of the variation in the product (such variation must usually be kept within limits, e.g. even though the average breaking strength of sewing thread is satisfactory, if the variation is great the public will object). Special tables have to be consulted for this purpose (see *Statistical Methods in Research and Production*, by O. L. Davies, Oliver and Boyd, Edinburgh).

Thus the theoretical work of many early mathematicians (Pascal, Laplace, Gauss) and of the mathematical statisticians of the last 50 years has been turned into great economic and industrial benefit, used throughout each day in every country of the world.

16

Some problems of practical origin

The problems and the data which now follow are drawn from actual practice. In some cases the entire original data are reproduced, but in others the data have been adjusted to give answers in 'round numbers' (these make useful examination questions). Certain of the problems illustrate the need for tests and results not given earlier and these are included in this chapter.

Examples

16.1. Natural sulphur is a mixture of four stable isotopes and so the atomic mass number of sulphur differs from one deposit to another, as different proportions of the isotopes are present in each. However, there is variation within the observations from one deposit some, or all, of which may be due to the method of measurement. The following 50 observations were made on a specimen of sulphur from a single deposit:

31·793 31·825 31·892 31·854 31·962 31·911 31·926 31·976 31·995
31·989 31·902 31·956 31·941 31·967 31·992 31·913 31·931 31·972
32·005 32·041 32·030 32·097 32·079 32·112 32·016 32·072 31·961
32·046 32·070 32·024 32·029 32·066 31·913 32·183 32·133 32·108
32·052 32·191 32·156 32·140 32·163 32·179 32·145 32·139 32·324
32·259 32·243 32·208 32·296 32·315

(a) Draw a histogram of these observations using groups (called box-widths in Chapter 2) of a convenient size.

(b) Assuming that the mean value of the observations in a single group is the mid-point of that group (e.g. observations in the group 32·000–32·050 have a mean 32·025) find the mean μ, the variance σ^2, and, correct to 3 significant figures, the standard deviation σ (divide by n and not $n - 1$ in computing σ^2, i.e. use (2.1) and (2.4)); this is the standard method when there are *many* observations, the use of (2.10) involves much arithmetic and greatly increases the chance of error, as may be found when part (d) is done).

(c) Improve the above estimate of σ^2 by subtracting $h^2/12$ from it, where h is the group (or 'box') width (neglect extreme groups); e.g. if the group-width is 0·050, we subtract $0·050^2/12 = 0·0002083$. ('Sheppard's Correction' as this is known allows approximately for the fact that the observations in a single group are *not* evenly spread,

135

in general. Unless there are some 2,000 or more readings it is usually smaller than errors due to randomness and so is usually omitted.)

(d) Calculate $\bar{x}$ and s by (2.9) and (2.10) and compare with μ and σ of (b).

(e) Taking the (b) values of μ and σ, draw a normal curve to fit your histogram (use Table A1).

(f) Find by (2.9) and (2.10) the mean and standard deviation of the following data obtained from another specimen of sulphur:

$$
\begin{array}{ccccc}
32{\cdot}062 & 32{\cdot}220 & 32{\cdot}136 & 32{\cdot}152 & 32{\cdot}121 \\
32{\cdot}041 & 32{\cdot}101 & 32{\cdot}187 & 32{\cdot}279 & 32{\cdot}202
\end{array}
$$

(g) Is there any reason to believe the specimens are from different deposits?

16.2. Fifty measurements of the solubility of potassium chloride in water at 60°C (expressed in g of solute per 100 g of water) were:

$$
\begin{array}{cccccccccc}
44{\cdot}96 & 45{\cdot}45 & 42{\cdot}82 & 46{\cdot}59 & 43{\cdot}35 & 45{\cdot}68 & 47{\cdot}13 & 46{\cdot}71 & 48{\cdot}20 & 44{\cdot}07 \\
48{\cdot}55 & 42{\cdot}20 & 45{\cdot}63 & 44{\cdot}49 & 46{\cdot}80 & 45{\cdot}32 & 44{\cdot}08 & 43{\cdot}19 & 45{\cdot}95 & 46{\cdot}86 \\
45{\cdot}96 & 44{\cdot}28 & 48{\cdot}51 & 45{\cdot}07 & 44{\cdot}20 & 45{\cdot}42 & 47{\cdot}73 & 45{\cdot}19 & 45{\cdot}91 & 46{\cdot}10 \\
45{\cdot}34 & 46{\cdot}59 & 44{\cdot}13 & 45{\cdot}77 & 43{\cdot}25 & 44{\cdot}56 & 48{\cdot}01 & 46{\cdot}07 & 46{\cdot}65 & 45{\cdot}66 \\
44{\cdot}81 & 47{\cdot}01 & 43{\cdot}62 & 45{\cdot}23 & 46{\cdot}99 & 45{\cdot}14 & 45{\cdot}37 & 43{\cdot}76 & 44{\cdot}48 & 45{\cdot}38
\end{array}
$$

Carry out the instructions (a), (b), (c), (d), and (e) of Example 16.1 on the above data and (f) on the following observations of the solubility of barium chloride:

$$
\begin{array}{cccccccccc}
45{\cdot}41 & 46{\cdot}41 & 45{\cdot}71 & 46{\cdot}49 & 46{\cdot}08 & 46{\cdot}70 & 46{\cdot}30 & 47{\cdot}09 & 46{\cdot}39 & 47{\cdot}42
\end{array}
$$

(g) Does barium chloride differ in solubility from potassium chloride?

16.3. Fifty determinations of the resistivity of platinoid (0·62 Cu, 0·15 Ni, 0·23 Zn) in $\mu\Omega$ cm were as follows:

$$
\begin{array}{cccccccccc}
33{\cdot}64 & 35{\cdot}87 & 36{\cdot}91 & 33{\cdot}23 & 37{\cdot}96 & 35{\cdot}81 & 34{\cdot}89 & 32{\cdot}30 & 34{\cdot}89 & 33{\cdot}76 \\
36{\cdot}33 & 33{\cdot}07 & 31{\cdot}80 & 34{\cdot}65 & 34{\cdot}62 & 34{\cdot}24 & 35{\cdot}49 & 33{\cdot}58 & 33{\cdot}09 & 32{\cdot}71 \\
35{\cdot}35 & 36{\cdot}51 & 33{\cdot}24 & 35{\cdot}83 & 32{\cdot}06 & 35{\cdot}40 & 33{\cdot}15 & 36{\cdot}59 & 35{\cdot}21 & 32{\cdot}02 \\
36{\cdot}18 & 34{\cdot}59 & 35{\cdot}27 & 34{\cdot}31 & 33{\cdot}12 & 34{\cdot}43 & 33{\cdot}04 & 35{\cdot}61 & 31{\cdot}27 & 34{\cdot}75 \\
33{\cdot}97 & 35{\cdot}59 & 34{\cdot}88 & 35{\cdot}14 & 37{\cdot}67 & 32{\cdot}32 & 34{\cdot}26 & 35{\cdot}70 & 34{\cdot}15 & 33{\cdot}03
\end{array}
$$

Carry out the instructions (a), (b), (c), (d), and (e) of 16.1 on this data and (f) on these measurements of the resistivity of another alloy:

$$
\begin{array}{cccccccccc}
33{\cdot}32 & 33{\cdot}71 & 34{\cdot}09 & 34{\cdot}18 & 34{\cdot}45 & 31{\cdot}46 & 31{\cdot}77 & 32{\cdot}04 & 32{\cdot}20 & 32{\cdot}78
\end{array}
$$

(g) Is this alloy unlikely to be platinoid?

16.4. Observations on the breaking strength of two specimens of yarn were:

Specimen	Number of Tests	Mean Strength x	Standard Deviation s
A	15	19·5 oz	3·1 oz
B	15	21·4 oz	4·2 oz

Is the difference in mean strength statistically significant?

16.5. The measurement of the amount of 'twist' in textile yarns is notoriously difficult and is subject to observer 'bias'. Two observers measure the twist at various points along the same yarn with these results:

Observer A   150 tests $\bar{x} = 7\cdot5$ turns per inch $s = 1\cdot0$ t.p.i.

Observer B   225 tests $\bar{x} = 7\cdot8$ turns per inch $s = 0\cdot7$ t.p.i.

Is there a significant difference between their mean estimates?

16.6. Measurements of the 'regain' (= moisture weight/dry weight) of two deliveries of 'wool tops' (i.e. wool which has been scoured, carded, and combed) are:

Delivery A, Percentage regain: 18·5 19·2 18·1 18·4 18·7 18·9
Delivery B, Percentage regain: 18·3 18·1 17·9 18·0 18·4 17·7

Is the difference in mean regain statistically significant?

Some problems involving χ^2 tests.

Examples

16.7. Fit a Poisson distribution (see Chapter 10) to the following data given by Yule and Greenwood, in the *Journal of the Royal Statistical Society*, 1920, on the number of accidents to each woman in a munitions factory:

Number of Accidents	0	1	2	3	4	5	6
Frequency (Number of Women)	447	132	42	21	3	2	0

Test the goodness of fit by the χ^2 test. If accidents happen at random then the fit should be good; if it is not what conclusion can be drawn?

16.8. Background radioactivity is usually random and the number of pulses observed in a small (10-second) interval should have a Poisson distribution. The following observations were taken in four laboratories A, B, C, and D. (i) Fit a Poisson distribution to each and test the goodness of fit. (ii) In the case of D, 100 readings were taken on one day and 100 on the following day. There was some reason to think the activity doubled on the second day. Fit a distribution to the data on this hypothesis and test the fit. (iii) Test the difference between the A and B, and (iv) between the C and D distributions.

Number of pulses per interval	0	1	2	3	4	5	6	7	8	9
Lab. A	2	8	15	19	20	16	10	6	3	1
Lab. B	5	15	21	23	18	10	5	2	1	0
Lab. C	22	33	19	8	9	6	2	1	0	0
Lab. D	27	50	42	32	19	16	8	3	2	1

N.B. If m_1, m_2 are the observed means and n_1, n_2 the total numbers of observations for two Poisson distributions, then, if

$$|m_1 - m_2|/\sqrt{(m_1/n_1 + m_2/n_2)} > 1\cdot96, 2\cdot58, 3\cdot29$$

the means differ significantly at, about, the 5, 1, 0·1 % level (because a Poisson variance = mean, and, with many readings, the mean is very nearly normally distributed, see 5.7).

16.9. In a dispute as to the relative merits of two colleges (see the London Sunday newspaper *Observer*, Feb. 21, 1965) the following figures of classes obtained by students in their final degree examination are quoted. Is there a significant difference between them?

Class:	1	upper 2	lower 2	3	Totals
College A	6	66	114	56	242
B	5	40	86	49	180

16.10. The following data has been taken (and regrouped slightly) from data given by Brownlee on the severity of smallpox attacks of vaccinated and unvaccinated persons. Is there significant difference between: (*a*) A, B, and C; (*b*) B and C; (*c*) C and A; (*d*) A and B?

Years since vaccination	Bad or very bad	Fairly bad	Mild	Totals
(A) 10	1	6	23	30
(B) 10–25	42	114	301	457
(C) Unvaccinated	65	41	9	115

16.11. In a Test Match series at cricket, the ratio of l.b.w. decisions to total number of wickets that fell is, for one side, 3:76 and, for the other, 9:52. (*a*) Are these significantly different? (*b*) Are there possible explanations other than that the umpires were biased?

16.12. Test the goodness of fit of the normal populations fitted in Examples (*a*) 16.1, (*b*) 16.2, (*c*) 16.3. Take $h = 0\cdot5\sigma$; find the observed frequencies between group boundaries . . . $\mu - 2h, \mu - h$, $\mu, \mu + h, \ldots$; for an observation *on* a boundary add 0·5 to each adjacent group frequency.

16.13. Draw a histogram and a frequency curve, and fit a normal distribution (finding μ and σ as in Example 16.1(*b*)) to the following six sets of data. Test the fit in each case (use Table A2 to find the proportions and, hence, the expected frequencies in each group).

(i) 200 determinations of g, the acceleration due to gravity, in cm/sec^2 are:

Group boundary	975	976	977	978	979	980	981	982	982	984	985	986	987
Frequency	1	2	8	21	28	46	34	32	12	8	7	1	

(ii) The number of hits between marking lines in 100 practice runs (along a given line of approach) by a squadron at bombing practice were (the distance D of marking line from target is in yards):

D	-220	-180	-140	-100	-60	-20	$+20$	$+60$	$+100$	$+140$	$+180$	$+220$	$+260$	$+300$
Frequency	1	2	6	9	8	15	18	14	11	8	4	2	2	

(iii) The results of measuring the diameter of fibres taken from a specimen of wool were as follows (a more detailed table than Table A2 is desirable here):

Diam. (to $\pm$ 1 micron)	8	10	12	14	16	18	20	22	24	26	28	30	32	34	36	38
Frequency	3	4	12	24	69	75	100	70	53	33	27	12	6	5	6	1

(iv) The breaking strength B of given lengths of a type of wool yarn:

B (g wt)	132	140	148	156	164	172	180	188	196	204	212	220	228	236
Frequency		1	5	10	19	34	32	39	38	31	20	13	5	1

(v) The diameters of the shells of calyptraea chinensis, a shellfish comparatively new to British shores, were measured by H. V. Wyatt (*Journal of Animal Ecology*, Nov. 1961) with these results (diameters in mm):

Diameter	5–	6–	7–	8–	9–	10–	11–	12–	13–	14–	15–	16–	17–	18–	19–	20–	21–
Frequency	11	15	36	52	64	72	44	36	40	31	40	26	29	14	3	2	1

(vi) Protons of a certain element were deflected by a magnetic field and then absorbed by a photographic film, forming a band of variable density. The number f of proton tracks in small strips of width 2×10^{-4} cm of this band were measured by graticule microscope at different distances x across the band, with these results:

x (mm)	0·0	0·25	0·50	0·75	1·00	1·25	1·50	1·75	2·00	2·25	2·50	2·75	3·00
f	2	9	51	148	309	343	291	352	305	137	58	11	1

(Note that f is the *ordinate* of the frequency curve here, and Table A1 must be used to find the expected freqs.; however μ and σ can be found just as before.)

(A) Is the mean in Example 16.13(ii) significantly different from zero?

16.14. In Example 16.13(v) and in (vi) it is suspected that the observations are the sum of two normal populations, these being possibly, in (v) the 1-year old shells and the 2- or more year old shells, and, in (vi) two isotopes of different atomic weight. Use the technique given below to fit two normal populations to (a) 16.13(v) and (b) to 16.13(vi), and test the fit in both cases.

To fit two normal populations to a set of observations. If the frequency curve fitted to a set of observations is bimodal (i.e. has maxima at two points or 'modes', as in Fig. 16.1) and if the χ^2 test

——————— Original frequency curves

— — — — — Reflection of outside half (shown thickened) of estimated constituent population

················ Constituent population estimated by subtraction

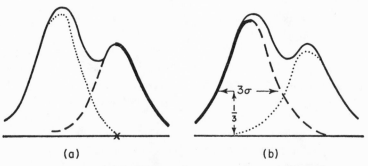

(a)   (b)

FIGURE 16.1. Fitting of two normal populations to a bimodal frequency curve

shows that a single normal population gives a bad fit, then possibly the observations come from two separate 'populations' (e.g. the heights of the adult population as a whole is bimodal, one mode being near the average man's height, the other near the average woman's).

The following graphical method can be used to fit two symmetrical constituent populations to the data. Take the mode with the smaller maximum (the right-hand one in Fig. 16.1(a), marked ×) and, taking that part of the frequency curve *outside* the mode (thickened in Fig. 16.1(a)), reflect it about the mode to get the broken curve. Subtract this broken curve from the original frequency curve to get the dotted curve of Fig. 16.1(a) which is a first estimate of the left-hand constituent population. Reflect the outside of this estimated population (thus getting the broken curve of Fig. 16.1(b)) and subtract from the frequency curve to obtain an estimate of the right-hand constituent population (dotted in Fig. 16.1(b)). We can reflect the outside of this and continue until no further improvement seems possible.

The method is based upon two facts, (i) a normal population is symmetrical, and (ii) it falls off rapidly with distance from its mean, hence the outside of one curve is scarcely affected by the other population.

The mean of each population is estimated by its mode, and its standard deviation from the fact that at a height of $\frac{1}{3}$ of maximum a normal curve has a width of 3σ (2.96σ is more exact), see Fig. 16.1(b). Note that if H_1, H_2 and σ_1, σ_2 are the maximum heights and standard deviations of the two populations the total frequency will be distributed between them in the ratio $H_1\sigma_1 : H_2\sigma_2$. The frequency contributed by each population to each group can then be calculated (use Table A2 or a more detailed table if available), and check the fit by the χ^2 test. Since the total expected and observed frequencies are equal and we fit *four* constants (two means and two standard deviations)

Degrees of freedom $\nu = $ No. of classes (groups) $- 5$.

Correlation and regression problems and some further tests.

Examples

16.15. The following were measurements made on shells of calyptraea chinensis (see 16.12), for specimens found (A) on stones, (B) inside paphia shells, (C) with clear growth rings:

(A) Height (mm)	y	4·1	4·2	4·3	4·7	4·5	4·5	5·0	5·5	6·2	4·6
Diameter (mm)	x	10·6	10·3	10·0	10·0	9·8	9·6	10·6	10·8	11·0	11·6

5·2	5·6	5·1	6·2	6·3	4·7	5·4	6·0	5·5	6·4	6·9	6·9	8·1

11·4	11·4	11·8	12·0	12·2	12·2	12·2	12·8	13·0	13·4	13·6	14·6	15·4

(B) Height (mm)	y	4·5	4·8	5·6	5·5	6·0	5·6	5·6	5·9	4·8
Diameter (mm)	x	10·6	12·6	13·8	15·0	15·6	17·0	17·2	17·4	14·0

(C) 1951 diameter	y	2·5	3·0	3·2	4·0	4·1	4·2	4·5	4·6	4·8
1952 diameter	x	8·1	9·1	9·7	9·6	10·7	11·5	11·4	11·3	10·5

	4·8	4·8	4·9	5·0	5·0	5·2	5·4	6·0	6·5

	10·5	11·9	11·4	11·4	12·6	12·7	13·5	13·1	13·9

(*a*) Calculate, for (A), (B), and (C), the correlation coefficient *r*, and the *y* on *x* and *x* on *y* regression lines. (*b*) Test whether the correlation and regression coefficients (i.e. the *y* on *x* and *x* on *y* slopes, see below) differ significantly from zero. (*c*) Test whether they differ from each other and (*d*) find their mean values, for A and B only. (*e*) Test whether observations (A) and (B) could have a common *y* on *x*, or a common *x* on *y* regression line (if A, B differ then environment affects growth).

To test a regression line slope (regression coefficient). Given a sample consisting of *n* pairs of observations $(x_1, y_1) \ldots (x_n, y_n)$, we first find $\bar{x}$, $\bar{y}$ the means of the *x*- and *y*-values separately, and then calculate

$$X = (x_1 - \bar{x})^2 + \ldots + (x_n - \bar{x})^2,$$

$$Y = (y_1 - \bar{y})^2 + \ldots + (y_n - \bar{y})^2,$$

$$C = (x_1 - \bar{x})(y_1 - \bar{y}) + \ldots + (x_n - \bar{x})(y_n - \bar{y})$$

Now C/X is the slope of the *y* on *x* regression line and is called the β_2 regression coefficient, see (13.11) *et seq.* Similarly C/Y is the β_1 regression coefficient. To test the hypothesis (I) that the true or population regression coefficient is β_2', (or (II) that the other coefficient is β_1'), we compute the test functions

$$t_2 = \frac{(C - \beta_2'X)\sqrt{(n-2)}}{\sqrt{(XY-C^2)}} \quad \text{or} \quad t_1 = \frac{(C - \beta_1'Y)\sqrt{(n-2)}}{\sqrt{(XY-C^2)}} \quad (16.1)$$

and if t_2 (or t_1) > two-tailed crit *t* ($v = n - 2$) we reject (I) (or II). To test whether the *y* on *x* regression slope differs from *zero*, put $\beta_2' = 0$ in (16.1) (note that if $\beta_2' = \beta_1' = 0$ then $t_2 = t_1$, so that the regression slopes are either both significantly different from zero or both not). The corresponding test for the correlation coefficient *r* is given by (13.14). N.B. we can calculate *X*, *Y*, and *C* more easily from the following alternative formulae:

$$X = x_1^2 + \ldots + x_n^2 - n\bar{x}^2$$

$$Y = y_1^2 + \ldots + y_n^2 - n\bar{y}^2 \quad (16.2)$$

$$C = x_1y_1 + \ldots + x_ny_n - n\bar{x}\bar{y}$$

To test regression line slopes (coefficients) from several samples. To test for significant differences (or 'variation') among the *y* on *x* slopes (i.e. regression coefficients β_2) from *k* samples consisting of

$n_1, n_2, \ldots n_k$ pairs of observations, we compute X, Y, and C for each sample, and then

$$S^2 = \frac{Y_1 + Y_2 + \ldots + Y_k - C_1^2/X_1 - C_2^2/X_2 - \ldots - C_k^2/X_k}{n_1 + n_2 + \ldots + n_k - 2k} \tag{16.3}$$

$$(S')^2 = \left(\frac{1}{k-1}\right)\left[\frac{C_1^2}{X_1} + \frac{C_2^2}{X_2} + \ldots + \frac{C_k^2}{X_k} - \frac{(C_1 + C_2 + \ldots + C_k)^2}{X_1 + \ldots + X_k}\right] \tag{16.4}$$

Finally, if $(S')^2/S^2 > \text{crit } F(v_1 = k - 1, v_2 = n_1 + \ldots + n_k - 2k)$ then we are confident that one y on x slope, at least, differs from the others. By replacing each Y_i by X_i and each X_i by Y_i in (16.3) and (16.4), we get the test for differences between the β_1 coefficients.

Formulae for mean regression coefficients (slopes) $\bar{\beta}_2$ and $\bar{\beta}_1$. Sometimes it is reasonable to assume that the y on x (or the x on y) slopes are the same in each sample and that it is only the *levels* of the regression lines that differ. The best or *mean* estimate of the two regression coefficients are then

$$\bar{\beta}_2 = \frac{C_1 + C_2 + \ldots + C_k}{X_1 + \ldots + X_k}, \quad \bar{\beta}_1 = \frac{C_1 + \ldots + C_k}{Y_1 + \ldots + Y_k} \tag{16.5}$$

To test the hypothesis that there is a common regression line. If $\bar{x}$, $\bar{y}$ are the *grand* means of *all* the x- and y-values, i.e. if

$$\bar{x} = \frac{n_1\bar{x}_1 + \ldots + n_k\bar{x}_k}{n_1 + \ldots + n_n}, \quad \bar{y} = \frac{n_1\bar{y}_1 + \ldots + n_k\bar{y}_k}{n_1 + \ldots + n_k}$$

compute

$$X_t = X_1 + n_1(\bar{x}_1 - \bar{x})^2 + \ldots + X_k + n_k(\bar{x}_k - \bar{x})^2,$$

$$Y_t = Y_1 + n_1(\bar{y}_1 - \bar{y})^2 + \ldots + Y_k + n_k(\bar{y}_k - \bar{y})^2$$

$$C_t = C_1 + n_1(\bar{x}_1 - \bar{x})(\bar{y}_1 - \bar{y}) + \ldots + C_k + n_k(\bar{x}_k - \bar{x})(\bar{y}_k - \bar{y}) \tag{16.6}$$

$$Z^2 = \frac{Y_t - C_t^2/X_t - (n_1 + n_2 + \ldots + n_k - 2k)S^2}{2k - 2} \tag{16.7}$$

If $Z^2/S^2 > \text{crit } F(v_1 = 2k - 2, v_2 = n_1 + \ldots + n_k - 2k)$ then we reject the common y on x regression line hypothesis. Replacing Y_t by X_i and X_i by Y_i in (16.3) and (16.7), we obtain the test for a common x on y line.

To test the levels for significant differences (assuming a common slope). If the ratio $(S')^2/S^2$, see (16.3) and (16.4), is small enough for it to be likely that the y on x regression lines from the k samples can all have a common slope, we may still wish to test whether the levels of these lines show significant variation. We then calculate.

$$(S'')^2 = \frac{Y_1 + \ldots + Y_k - (C_1 + \ldots + C_k)^2/(X_1 + \ldots + X_k)}{n_1 + \ldots + n_k - k - 1}$$

(16.8)

$$(Z')^2 = \frac{Y_t - C_t^2/X_t - (n_1 + \ldots + n_k - k - 1)(S'')^2}{k - 1}$$

(16.9)

If $(Z')^2/(S'')^2 > \text{crit } F(v_1 - k = 1, v_2 = n_1 + \ldots + n_k - k - 1)$ we are confident that the levels have significant differences.

N.B. X_t, Y_t, and C_t are in fact the values of X, Y, and C calculated by *all* the pairs of observations (x_j, y_j) as belonging to a single sample; hence

$$\bar{x} = (\textstyle\sum x_j)/N, \quad \bar{y} = (\textstyle\sum y_j)/N, \quad X_t = \textstyle\sum x_j^2 - N\bar{x}^2,$$
$$Y_t = \textstyle\sum y_j^2 - N\bar{y}^2, \quad C_t = \textstyle\sum x_j y_j - N\bar{x}\bar{y}$$

(16.10)

where $N = n_1 + \ldots + n_k$ and j goes from 1 to N in each summation $\sum$. (All the above tests are deduced on this basis: We fit lines in two ways; if the variance of the 'errors' one way is significantly greater than the other, the second is the better way. We assume *only* that the errors are distributed normally and, for example, do not restrict the underlying populations to the normal bivariate.)

To find the mean correlation coefficient. For a single sample the estimated correlation coefficient $r = C/\sqrt{(XY)}$, see (13.4). If we assume that several samples come from populations with the *same* correlation coefficient ρ, then the best or *mean* estimate $\bar{r}$ of ρ is found thus. First find the Z function for each sample (N.B. Fisher's z, see (13.6) $= 1\cdot1513Z$), where

$$Z_i = \log_{10}\{(1 + r_i)/(1 - r_i)\}$$

Then find $\bar{Z}$, the mean Z, and from it $\bar{r}$ by the formulae

$$\bar{Z} = \frac{(n_1 - 3)Z_1 + (n_2 - 3)Z_2 + \ldots + (n_k - 3)Z_k}{(n_1 + \ldots + n_k - 3k)}$$

(16.11)

$$\bar{r} = \tanh \bar{z} = \frac{\text{antilog }(\bar{Z}) - 1}{\text{antilog }(\bar{Z}) + 1}$$

(16.12)

To test several correlation coefficients for significant difference.
Compute the test function U where (N.B. $1 \cdot 1513^2 = 1 \cdot 3255$),

$$U = 1 \cdot 3255\{(n_1 - 3)(Z_1 - \bar{Z})^2 + \ldots + (n_k - 3)(Z_k - \bar{Z})^2\}$$
(16.13)

and if this $> \chi^2_{0 \cdot 05}(\nu = k - 1)$ then we are confident that there is significant difference between some, at least, of the coefficients (at the 5% level). If $k = 2$ an alternative but *exactly* equivalent test is to find whether

$$|Z_1 - Z_2|/\sqrt{\{1/(n_1 - 3) + 1/(n_2 - 3)\}} > 1 \cdot 96/1 \cdot 1513 = 1 \cdot 702.$$

(The correlation tests are based on the fact that for random samples the function z is almost normally distributed with a variance of $1/(n - 3)$.)

Examples

16.16. Calculate the correlation coefficient, the y on x, and the x on y regression lines for each of these 3 sets of data on the crushing strength (in $lb/in^2 \div 10$) of mortar and concrete made from 7 specimens of cement (see the paper by Glanville and Reid, *Structural Engineer*, 1934) after various times of setting:

(A) after	Concrete, y	138	278	49	293	104	350	141
1 day:	Mortar, x	263	493	137	477	233	568	230
(B) after	Concrete, y	507	653	425	662	437	735	439
7 days:	Mortar, x	750	936	453	893	545	797	631
(C) after	Concrete, y	679	818	651	842	603	832	724
28 days:	Mortar, x	895	1066	632	1100	716	897	846

(*a*) Test the correlation and regression coefficients for significant difference from zero, (*b*) find their mean values, and (*c*) test for significant differences between them. Test the hypotheses that there is (*d*) a common y on x, (*e*) a common x on y regression line, (*f*) that the y on x levels, (*g*) the x on y levels differ significantly. (If the y on x slopes, i.e. the regression coefficients β_2 do not differ significantly, then tests after 1, instead of 28 days, can be used, saving time and money.)

16.17. The following measurements of yarn breaking strength y against percentage humidity x were made on a specimen of cotton yarn:

y	102 107 113 101 98 106 98 107 124 116 118 112 103 117 97 96 123 104 112
x	73 44 77 56 46 59 56 64 74 56 88 65 58 92 58 44 84 47 53

Find the correlation coefficient, the y on x and x on y regression lines. Test the correlation and regression coefficients for significant difference from zero. (The relation between breaking strength and humidity has played an important part in determining the location of the textile industry.) Is y better fitted by a quadratic curve (see Sections 12.3 and 12.6)?

16.18. The following determinations of the specific heat y of solder (0·54 tin, 0·46 lead) in mcal/g deg C were made at various temperatures x:

y	44·75	44·44	45·25	44·82	43·74	43·81	44·41	44·26	44·09	43·81	43·78	44·68
x	52·5	48·0	47·5	46·0	44·0	43·0	42·5	42·5	42·0	41·0	39·5	38·5

y	43·14	43·70	43·25	44·14	43·72	43·32	44·34	43·41	43·24	43·69	43·76	42·70	42·25
x	38·0	37·5	37·0	36·5	35·5	34·5	33·5	33·0	32·5	32·0	31·0	29·0	27·5

Find the correlation coefficient and y on x regression line. Test whether the correlation and β_2 regression coefficients differ significantly from zero. (Most specific heats show a slight change with temperature representable as a straight line to a good approximation.) Are the specific heat observations better fitted by a quadratic curve in x (see Sections 12.3 and 12.6)?

Some least squares line fitting.

Examples

16.19. The activity a of a radioactive substance at a time t is given by the formula $a = a_0 e^{-\lambda t}$. If we take logs this becomes

$$\log_e a = \log_e a_0 - \lambda t \tag{16.14}$$

(note that the Napierian logarithm $\log_e x = 0·4343 \log_{10} x$ where $\log_{10}$ is the common logarithm; a useful result if tables of $\log_e$ are not available). Observations with a Geiger counter on radioactive titanium $^{45}_{22}$Ti gave the results:

Time in hours t	0	1	2	3	4
Activity in pulses per min a	3903	2961	2463	1912	1553

Plot $\log_e a$ against t and fit the best straight line. The slope of this line is the best estimate of λ. Find the half-life (i.e. the time for the activity a to fall from a_0 to $0·5a_0$) and find the expected activity after 8 hours for an initial activity of 4,000.

16.20. Carry out the instructions of Example 16.19 on these observations of the radioactivity of a specimen of ^{116}In:

t	0	10	20	30	40	50	60
a	1960	1792	1536	1380	1188	1081	904

16.21. Newton's law of cooling leads to the relation

$$\theta = \theta_0 \, e^{-kt}$$

where θ is the excess temperature of a body over its surroundings at time t, and k depends on the nature of the body. Observations on a calorimeter containing, initially, a hot liquid were

t (min)	0	10	20	30	40	50	60
θ (degC)	50·0	36·7	27·8	20·0	14·8	11·3	8·1

(a) Find the best value of k, (b) calculate the time of cooling from 40° to 20° and (c) from 30° to 15°C.

Rank correlation (relation to correlation coefficient; ties). There is a rather approximate relation between the (true or population) correlation coefficient ρ and Spearman's rank correlation coefficient ρ_s, see Section 13.7, namely

$$\rho = 2 \sin (\pi \rho_s/6) = 2 \sin (30 \rho_s{}^\circ) \qquad (16.15)$$

If there are 'ties', that is if several observations are ranked equal then the total value of the ranks which would be assigned to them is averaged amongst them (in Example 16.22 the cloths G_2 and H_2 are ranked equal by touch; instead of arbitrarily assigning one the rank value 3 and the other 4, we given them both 3·5). *However*, the test for significance (13.7.3) is then affected, so less reliance can be placed on borderline significance results in such cases.

Examples

16.22. Binns in a paper in the *Journal of the Textile Institute*, Vol. 25, T89, gave the following results of the ranking of ten worsted cloths by skilled assessors, on the basis of *sight, touch*, and *full* (i.e. that which they would buy if all were offered at the same price):

Cloth Number	F	G_1	H_1	J_1	L_1	G_2	M_1	H_2	P	Q
Sight	4	3	6	1	2	5	8	7	9	10
Touch	2	7	1	6	5	3·5	9	3·5	10	8
Full	1	2	3	4	5	6	7	8	9	10

Find Spearman's ρ_s for (i) sight and touch, (ii) sight and full, and (iii) touch and full and test for significance.

16.23. Calculate ρ_s, test it for significant difference from zero, and estimate ρ from 16.15 for each set of data in Examples 16.15, 16.16, 16.17, and 16.18.

Estimating errors in direct and indirect measurement. Repeated measurements of a quantity always show slight variations or 'errors'. (these are *not* mistakes but are due to the impossibility of reproducing highly accurate measurements exactly, or may be due to variation in the quantity itself). The standard deviation of these measurements, calculated in the usual way (2.10), is a measure of the accuracy or repeatability that can be obtained by the method of measurement employed, and so is a very useful thing to know. Now, some quantities are the sum or difference of directly measured other quantities (e.g. in Example 16.25 the wire diameter is the difference of two micrometer readings). Suppose that $u = x \pm y \pm z \ldots$ then the variance of the 'error' in estimating u is the sum of the variances in measuring $x, y, \ldots$, (see 5.15) provided the errors in one measurement do not affect the others. That is

$$\text{var } u = \text{var } x + \text{var } y + \text{var } z + \ldots \qquad (16.16)$$

There are many other quantities that can be expressed as a function of directly measurable quantities (see Example 16.24) but cannot be easily measured directly. Suppose U is such a quantity and that $U = f(x, y, z, \ldots)$. It can be proved that provided the variances of $x, y, \ldots$ are small, then

$$\text{var } U = \left(\frac{\partial f}{\partial x}\right)^2 \text{var } x + \left(\frac{\partial f}{\partial y}\right)^2 \text{var } y + \left(\frac{\partial f}{\partial z}\right)^2 \text{var } z + \ldots \qquad (16.17)$$

where, in calculating $\partial f/\partial x$, $\partial f/\partial y$, ... we use the mean values of $x, y, \ldots$ The standard deviation $S = \sqrt{(\text{var } U)}$ gives a measure of the accuracy of the estimation of U in the same units as U itself. S is called the standard error of U (see Section 5.8), but older scientific text-books often quote the 'probable error' which $\doteqdot 0.6745S$ (when there are many observations), however, the term is scarcely used nowadays.

16.24. Now g the acceleration due to gravity can be measured by the simple pendulum experiment where if L is the pendulum length and T the time of one complete vibration

$$T = 2\pi\sqrt{(L/g)}, \text{ that is } g = T^2/(4\pi^2 L) \qquad (16.18)$$

Ten measurements of L in cm and ten measurements of T in seconds were:

L	106·42	106·43	106·44	106·45	106·45	106·46	106·46	106·46	106·46	106·47
T	103·6	103·4	103·4	104·0	103·0	103·6	103·2	103·4	103·8	103·6

Find (i) the formula for var g in terms of var T and var L; (ii) find var T and var L and determine $S = \sqrt{(\text{var } g)}$. Assuming that measurements of T and L take equal time and money, what is the desirable ratio of numbers of L to numbers of T measurements? It can be shown that, other things being equal, it is best to arrange the experimental method so that, in the notation of (16.17)

$$\left(\frac{\partial f}{\partial x}\right)^2 \text{var } x = \left(\frac{\partial f}{\partial y}\right)^2 \text{var } y = \left(\frac{\partial f}{\partial z}\right)^2 \text{var } z = \ldots \quad (16.19)$$

16.25. The density of a specimen of copper in wire form is to be found. The following 10 values in cm for the 'zero reading' of a micrometer screw-gauge were obtained:

0·003 0·002 0·003 0·002 0·001 0·003 0·001 0·001 0·001 0·001

and the following 10 micrometer readings for the wire diameter:

0·087 0·084 0·088 0·082 0·085 0·086 0·087 0·083 0·086 0·082

The mean measurement m and the standard deviation S were (i) 55·22, 0·075 cm for the length of the wire, and (ii) 2·785, 0·004 g for the mass of the wire. Find m and S for (a) the zero reading, (b) the micrometer reading of wire diameter, (c) the estimate of wire diameter, itself, (d) for the wire radius, (e) for the wire cross-section, (f) for the volume of the wire, and (g) for the density of copper.

16.26. With the same micrometer screw-gauge as in Example 16.25, the density of lead was determined from the following 10 measurements of the diameter of a lead sphere, whose mass was accurately determined to be 47·333 g:

2·002 2·003 2·000 2·002 2·001 2·001 2·002 2·002 2·004 2·003

Find the mean value m and the standard deviation S of the estimates of (a) sphere diameter, (b) sphere volume, (c) the density of lead.

16.27. The refractive index η of a substance is given by

$$\eta = \sin i / \sin r$$

The mean m and standard deviation S of i were 49° 30′, 0° 30′, and of r were 30° 0′, 0° 50′, (a) give the formula for var η in terms of var i and var r, and, (b), find m and S for the estimate of η.

Some analysis of variance problems.

Examples

16.28. The mean frequency of breakdown, averaged over two months, of the electronic digital computers used by a large organization, at different times of the day are as follows. Is there significant difference (a) between computers, (b) between times of day?

Time of day	9–11	11–13	13–15	15–17
A .	0·4	0·6	0·4	1·0
Computer B .	0·6	1·2	1·4	1·6
C .	0·2	0·6	1·2	0·4

16.29. The yield of a product made with six different commercial specimens of acid, and estimated by five different types of test are as follows. Is (a) the type of acid, (b) the type of test, causing significant variation in the results?

Acid number	1	2	3	4	5	6	Totals	
1 .	93	109	101	93	96	102	594	
Test 2 .	88	99	104	89	107	95	582	
Number: 3 .	104	108	109	93	113	97	624	$\Sigma x_{ij}^2 = 313,446$
4 .	104	106	111	104	109	96	630	
5 .	106	103	105	111	105	100	630	
	490	520	525	485	525	485	3060	

16.30. The following data, taken from a paper by Warren (*Biometrika*, 1909), gives measurements of termite headbreadths from various nests during various months. Is there significant variation (a) from nest to nest, (b) from month to month? The figure tabulated is 1,000 × (headbreadth in mm − 2).

Nest number	668	670	672	674	675	Totals	
Nov	273	479	404	447	456	2,059	
Jan	332	603	457	388	626	2,406	$\Sigma x_{ij}^2 =$
Mar	375	613	452	515	633	2,588	5,114,571
May	373	557	396	445	487	2,258	
Aug	318	377	279	312	410	1,696	
Totals	1,671	2,629	1,988	2,107	2,612	11,007	

Appendix

Note on algebra

(*a*) **The number of combinations.** Suppose we have four coloured balls: red, orange, yellow, and blue, and we wish to find the number of 'combinations' (= different selections) taking two balls at a time. The different selections here are RO, RY, RB, OY, OB, YB and so the number of combinations is 6 (note that OR is not a different selection from RO since the *order* does not matter in a selection).

Examples

A(i). Enumerate the different selections of 5 coloured balls, R, O, Y, G, B say, taken 2, 3, or 4 at a time, and give the number of combinations in each case.

It can be proved mathematically that the number of combinations of n things (all different from each other) taken s at a time is

$$^nC_s = \frac{n(n-1)(n-2)\ldots(n-s+1)}{1 \times 2 \times 3 \times \ldots \times s} \tag{A.1}$$

If we put $s = 1$ in formula (A.1) we see that $^nC_1 = n$, and putting $s = n$ we see that $^nC_n = 1$, whatever the value of n, a result to remember. Note, too, that $1 \times 2 \times 3 \times \ldots \times (s-1)s$ is written $s!$ and called 'factorial s'. Thus $3! = 1 \times 2 \times 3 = 6$, and $4! = 24$.

A(ii). Calculate $^5C_2$, $^5C_3$, $^5C_4$ and check your answers with Example A(i).

A(iii). Calculate $^7C_3$, $^{10}C_2$, $^9C_1$, $^8C_4$.

A(iv). Calculate $5!$, $7!$, $8!$.

The following is a very useful relation:

$$^nC_{n-s} = {}^nC_s \tag{A.2}$$

Thus, for example, if $n = 5$, $s = 4$ we have that $^5C_4 = {}^5C_1$; the former is quite lengthy but the latter is simple (it is 5). If we put $s = n$ in (A.2) then we have the important result $^nC_0 = {}^nC_n = 1$. Collecting the special cases $s = 0, 1$, and n together we have

$$^nC_0 = 1; \quad {}^nC_1 = n; \quad {}^nC_n = 1 \tag{A.3}$$

A(v). Find $^6C_0$, $^6C_2$, $^6C_4$, $^6C_5$, $^{10}C_2$, $^{10}C_7$, $^{10}C_{10}$, $^{12}C_{11}$.

(b) **The binomial theorem.** It can be proved that if n is a positive integer (= whole number) then the following relation holds whatever the values of p and q:

$$(p + q)^n = p^n + {}^nC_1 p^{n-1}q + {}^nC_2 p^{n-2}q^2 + \cdots$$
$$+ {}^nC_{n-2} p^2 q^{n-2} + {}^nC_{n-1} pq^{n-1} + {}^nC_n q^n \quad (A.4)$$

A(vi). Use (A.4) to write out the expansion of $(p + q)^3$, $(p + q)^5$, $(a + b)^4$.

Tables

Acknowledgments. Tables A.3 and A.6 have been condensed from *Biometrika Tables for Statisticians*, Cambridge University Press, and Table A.4 from the original papers, namely ASPIN, A. A. 'Tables for Use in Comparisons . . .' *Biometrika* (1949), **36**, 290, and W. H. TRICKETT, B. L. WELCH, and G. S. JAMES 'Further Critical Values for the Two-Means Problem' *Biometrika*, (1956), **43**, 203, by permission of the *Biometrika* Trustees and the authors.

Table A.5 has been condensed from R. FISHER, *Statistical Methods for Research Workers* by permission of the publishers Oliver and Boyd, Edinburgh.

The reader's attention is drawn to these tables and to *Statistical Tables* by Sir Ronald Fisher and F. Yates, Oliver and Boyd, and to *The Methods of Statistics* by L. H. C. Tippett, Williams and Norgate, London which also gives charts enabling *any* probability level to be used (and not just the 5, 1, or 0·1 % levels as with most tables).

Interpolation. Interpolation, or the determining of critical values for those degrees of freedom which are not given explicitly in the tables is discussed in Subsection 6.10(c).

Table A1. VALUES OF $\sigma p(x)$ FOR THE NORMAL DISTRIBUTION

Note: x is the variate-value; σ is the population standard deviation; $p(x)$ is the probability density (which is symmetrical about $x = \mu$, i.e. $p(\mu + h) = p(\mu - h)$).

$\lvert x - \mu \rvert / \sigma$	0	0·1	0·2	0·3	0·4	0·5	0·6	0·7	0·8
$\sigma p(x)$	0·3989	0·3970	0·3910	0·3814	0·3683	0·3521	0·3332	0·3123	0·2897

$\lvert x - \mu \rvert / \sigma$	0·9	1·0	1·2	1·4	1·6	1·8	2·0	2·2	2·4
$\sigma p(x)$	0·2661	0·2420	0·1942	0·1497	0·1109	0·0790	0·0540	0·0355	0·0224

$\lvert x - \mu \rvert / \sigma$	2·6	2·8	3·0	3·2	3·4	3·6	3·8	4·0	
$\sigma p(x)$	0·0136	0·0079	0·0044	0·0024	0·0012	0·0006	0·0003	0·0001	

Table A2. Probabilities Associated with 'Tails' of the Normal Distribution

Note: Probabilities in the left-hand tail = Probabilities in the right-hand tail, i.e. $\Pr(x - \mu \leqslant -k\sigma) = \Pr(x - \mu \geqslant k\sigma)$

k	0	0·1	0·2	0·3	0·4	0·5	0·6	0·7
$\Pr(x - \mu \geqslant k\sigma)$	0·5000	0·4602	0·4207	0·3821	0·3446	0·3085	0·2743	0·2420

k	0·8	0·9	1·0	1·2	1·4	1·6	1·8	2·0
$\Pr(x - \mu \geqslant k\sigma)$	0·2119	0·1841	0·1587	0·1151	0·0808	0·0548	0·0359	0·0228

k	2·2	2·4	2·6	2·8	3·0	3·2	3·4	3·6
$\Pr(x - \mu \geqslant k\sigma)$	0·0139	0·00820	0·00466	0·00256	0·00135	0·00069	0·00034	0·00016

k	3·8	4·0	4·2
$\Pr(x - \mu \geqslant k\sigma)$	0·00007	0·00003	0·00001

Table A3. Critical Values of t

Probability Level (%)	10	5	1	0·1	Two-tailed	10	5	1	0·1
	5	2·5	0·5	0·05	One-tailed	5	2·5	0·5	0·05
1	6·31	12·7	63·7	637	10	1·81	2·23	3·17	4·59
Degrees 2	2·92	4·30	9·92	31·6	12	1·78	2·18	3·05	4·32
of 3	2·35	3·18	5·84	12·9	15	1·75	2·13	2·95	4·07
freedom 4	2·13	2·78	4·60	8·61	ν: 20	1·72	2·09	2·85	3·85
ν: 5	2·01	2·57	4·03	6·86	25	1·71	2·06	2·79	3·72
6	1·94	2·45	3·71	5·96	30	1·70	2·04	2·75	3·65
7	1·89	2·36	3·50	5·40	40	1·68	2·02	2·70	3·55
8	1·86	2·31	3·36	5·04	60	1·67	2·00	2·66	3·46
9	1·83	2·26	3·25	4·78	∞	1·64	1·96	2·58	3·29

Table A4. CRITICAL VALUES OF W (TWO-TAILED) OR W' (ONE-TAILED) AT VARIOUS PROBABILITY LEVELS

Notes:

$$W = \frac{|\bar{x} - \bar{x}'|}{\sqrt{[s^2/n + s'^2/n']}}; \quad h = \frac{s^2/n}{s^2/n + s'^2/n'}$$

we write $v = n - 1$, $v' = n' - 1$, and take v to be the larger, i.e.
$v \geqslant v'$; interpolation with respect to $1/v$ and/or $1/v'$ gives greater
accuracy.

Level		10 (for W), 5 (for W') %							5 (for W), 2·5 (for W') %						
h		0·0	0·2	0·4	0·5	0·6	0·8	1·0	0·0	0·2	0·4	0·5	0·6	0·8	1·0
v'	v														
6	6	1·94	1·85	1·76	1·74	1·76	1·85	1·94							
	8	1·94	1·85	1·76	1·73	1·74	1·79	1·86							
	10	1·94	1·85	1·76	1·73	1·73	1·76	1·81			No figures yet computed				
	20	1·94	1·85	1·76	1·73	1·71	1·70	1·72							
	∞	1·94	1·85	1·76	1·72	1·69	1·66	1·64							
8	8	1·86	1·79	1·73	1·73	1·73	1·79	1·86	2·31	2·20	2·10	2·08	2·10	2·20	2·31
	10	1·86	1·79	1·73	1·72	1·72	1·76	1·81	2·31	2·20	2·10	2·08	2·08	2·14	2·23
	20	1·86	1·79	1·73	1·71	1·70	1·70	1·72	2·31	2·20	2·10	2·06	2·04	2·05	2·09
	∞	1·86	1·79	1·72	1·70	1·68	1·65	1·64	2·31	2·20	2·09	2·05	2·01	1·97	1·96
10	10	1·81	1·76	1·72	1·71	1·72	1·76	1·81	2·23	2·14	2·08	2·06	2·08	2·14	2·23
	20	1·81	1·76	1·71	1·70	1·69	1·70	1·72	2·23	2·14	2·07	2·05	2·04	2·05	2·09
	∞	1·81	1·76	1·71	1·69	1·67	1·65	1·64	2·23	2·14	2·06	2·03	2·00	1·97	1·96
20	20	1·72	1·70	1·68	1·68	1·68	1·70	1·72	2·09	2·05	2·02	2·02	2·02	2·05	2·09
	∞	1·72	1·70	1·67	1·66	1·66	1·65	1·64	2·09	2·04	2·01	1·99	1·97	1·96	1·96
∞	∞	1·64	1·64	1·64	1·64	1·64	1·64	1·64	1·96	1·96	1·96	1·96	1·96	1·96	1·96

Table A4 *(contd.)*

Level		2 (for W), 1 (for W')%							1 (for W), 0·5 (for W')%						
h		0·0	0·2	0·4	0·5	0·6	0·8	1·0	0·0	0·2	0·4	0·5	0·6	0·8	1·0
v'	v														
10	10	2·76	2·63	2·51	2·50	2·51	2·63	2·76	3·17	3·00	2·82	2·79	2·82	3·00	3·17
	12	2·76	2·63	2·51	2·49	2·49	2·57	2·68	3·17	3·00	2·82	2·78	2·79	2·91	3·05
	20	2·76	2·63	2·51	2·47	2·45	2·47	2·53	3·17	3·00	2·82	2·76	2·73	2·76	2·85
	∞	2·76	2·63	2·50	2·44	2·40	2·34	2·33	3·17	2·99	2·82	2·74	2·67	2·60	2·58
12	12	2·68	2·57	2·48	2·47	2·48	2·57	2·68	3·05	2·91	2·78	2·76	2·78	2·91	3·05
	20	2·68	2·57	2·48	2·45	2·44	2·47	2·53	3·05	2·91	2·78	2·74	2·72	2·76	2·85
	∞	2·68	2·57	2·46	2·42	2·38	2·34	2·33	3·05	2·91	2·77	2·71	2·65	2·59	2·58
20	20	2·53	2·46	2·42	2·42	2·42	2·46	2·53	2·85	2·76	2·70	2·70	2·70	2·76	2·85
	∞	2·53	2·46	2·40	2·38	2·36	2·33	2·33	2·85	2·76	2·68	2·65	2·62	2·59	2·58
∞	∞	2·33	2·33	2·33	2·33	2·33	2·33	2·33	2·58	2·58	2·58	2·58	2·58	2·58	2·58

A good approximate formula which agrees well with the above values of W, W' and can also be used for v, v' and probability levels outside the above ranges is this
crit W = (two-tailed) crit t (degrees of freedom = δ), crit W' = one-tailed t
where W, or W' and t are both taken at the same probability level and

$$\delta = \frac{1}{h^2/v + (1-h)^2/v'}$$

Table A.5. Critical Values of χ^2

Note: For $\nu > 100$, use the formula $\chi^2 = 0 \cdot 5\{\beta + \sqrt{(2\nu - 1)}\}^2$, the values of β are in the last column below

ν	1	2	3	4	5	6	7	8	9	10	12
$\chi^2_{0\cdot995}$	0·000	0·010	0·072	0·207	0·412	0·676	0·989	1·34	1·73	2·16	3·07
$\chi^2_{0\cdot99}$	0·000	0·020	0·115	0·297	0·554	0·872	1·24	1·65	2·09	2·56	3·57
$\chi^2_{0\cdot975}$	0·000	0·051	0·216	0·484	0·831	1·24	1·69	2·18	2·70	3·25	4·40
$\chi^2_{0\cdot95}$	0·000	0·103	0·352	0·711	1·15	1·64	2·17	2·73	3·33	3·94	5·23
$\chi^2_{0\cdot05}$	3·84	5·99	7·81	9·49	11·1	12·6	14·1	15·5	16·9	18·3	21·0
$\chi^2_{0\cdot025}$	5·02	7·38	9·35	11·1	12·8	14·4	16·0	17·5	19·0	20·5	23·3
$\chi^2_{0\cdot01}$	6·63	9·21	11·3	13·3	15·1	16·8	18·5	20·1	21·7	23·2	26·2
$\chi^2_{0\cdot005}$	7·88	10·6	12·8	14·9	16·7	18·5	20·3	22·0	23·6	25·2	28·3
$\chi^2_{0\cdot001}$	10·8	13·8	16·3	18·5	20·5	22·5	24·3	26·1	27·9	29·6	32·9

ν	14	16	18	20	22	24	26	28	30	40	60	100	β
$\chi^2_{0\cdot995}$	4·07	5·14	6·26	7·43	8·64	9·89	11·2	12·5	13·8	20·7	35·5	67·3	−2·58
$\chi^2_{0\cdot99}$	4·66	5·81	7·01	8·26	9·54	10·9	12·2	13·6	15·0	22·2	37·5	70·1	−2·33
$\chi^2_{0\cdot975}$	5·63	6·91	8·23	9·59	11·0	12·4	13·8	15·3	16·8	24·4	40·5	74·2	−1·96
$\chi^2_{0\cdot95}$	6·57	7·96	9·39	10·9	12·3	13·8	15·4	16·9	18·5	26·5	43·2	77·9	−1·64
$\chi^2_{0\cdot05}$	23·7	26·3	28·9	31·4	33·9	36·4	38·9	41·3	43·8	55·8	79·1	124·3	+1·64
$\chi^2_{0\cdot025}$	26·1	28·8	31·5	34·2	36·8	39·4	41·9	44·5	47·0	59·3	83·3	129·6	+1·96
$\chi^2_{0\cdot01}$	29·1	32·0	34·8	37·6	40·3	43·0	45·6	48·3	50·9	63·7	88·4	135·8	+2·33
$\chi^2_{0\cdot005}$	31·3	34·3	37·2	40·0	42·8	45·6	48·3	51·0	53·7	66·8	92·0	140·2	+2·58
$\chi^2_{0\cdot001}$	36·1	39·3	42·3	45·3	48·3	51·2	54·1	56·9	59·7	73·4	99·6	149·4	+3·09

Table A6. CRITICAL VALUES OF F AT THE 5 AND 1% PROBABILITY
LEVELS

Note: Interpolation with respect to $1/v_1$ and/or $1/v_2$ gives greater accuracy.

5% *values*

v_1	1	2	3	4	5	6	8	10	12	15	20	30	60	∞
v_2														
1	161	199	216	225	230	234	239	242	244	246	248	250	252	254
2	18·5	19·0	19·2	19·3	19·3	19·3	19·4	19·4	19·4	19·4	19·5	19·5	19·5	19·5
3	10·1	9·55	9·28	9·12	9·01	8·94	8·85	8·79	8·74	8·70	8·66	8·62	8·57	8·53
4	7·71	6·94	6·59	6·39	6·26	6·16	6·04	5·96	5·91	5·86	5·80	5·75	5·69	5·63
5	6·61	5·79	5·41	5·19	5·05	4·95	4·82	4·74	4·68	4·62	4·56	4·50	4·43	4·36
6	5·99	5·14	4·76	4·53	4·39	4·28	4·15	4·06	4·00	3·94	3·87	3·81	3·74	3·67
8	5·32	4·46	4·07	3·84	3·69	3·58	3·44	3·35	3·28	3·22	3·15	3·08	3·01	2·93
10	4·96	4·10	3·71	3·48	3·33	3·22	3·07	2·98	2·91	2·85	2·77	2·70	2·62	2·54
12	4·75	3·89	3·49	3·26	3·11	3·00	2·85	2·75	2·69	2·62	2·54	2·47	2·38	2·30
15	4·54	3·68	3·29	3·06	2·90	2·79	2·64	2·54	2·48	2·40	2·33	2·25	2·16	2·07
20	4·35	3·49	3·10	2·87	2·71	2·60	2·45	2·35	2·28	2·20	2·12	2·04	1·95	1·84
30	4·17	3·32	2·92	2·69	2·53	2·42	2·27	2·16	2·09	2·01	1·93	1·84	1·74	1·62
60	4·00	3·15	2·76	2·53	2·37	2·25	2·10	1·99	1·92	1·84	1·75	1·65	1·53	1·39
∞	3·84	3·00	2·60	2·37	2·21	2·10	1·94	1·83	1·75	1·67	1·57	1·46	1·32	1·00

1% *values*

v_1	1	2	3	4	5	6	8	10	12	15	20	30	60	∞
v_2														
1	4052	4999	5403	5625	5764	5859	5982	6056	6106	6157	6209	6261	6313	6366
2	98·5	99·0	99·2	99·2	99·3	99·3	99·4	99·4	99·4	99·4	99·4	99·5	99·5	99·5
3	34·1	30·8	29·5	28·7	28·2	27·9	27·5	27·2	27·1	26·9	26·7	26·5	26·3	26·1
4	21·2	18·0	16·7	16·0	15·5	15·2	14·8	14·5	14·4	14·2	14·0	13·8	13·7	13·5
5	16·3	13·3	12·1	11·4	11·0	10·7	10·3	10·1	9·89	9·72	9·55	9·38	9·20	9·02
6	13·7	10·9	9·78	9·15	8·75	8·47	8·10	7·87	7·72	7·56	7·40	7·23	7·06	6·88
8	11·3	8·65	7·59	7·01	6·63	6·37	6·03	5·81	5·67	5·52	5·36	5·20	5·03	4·86
10	10·0	7·56	6·55	5·99	5·64	5·39	5·06	4·85	4·71	4·56	4·41	4·25	4·08	3·91
12	9·33	6·93	5·95	5·41	5·06	4·82	4·50	4·30	4·16	4·01	3·86	3·70	3·54	3·36
15	8·68	6·36	5·42	4·89	4·56	4·32	4·00	3·80	3·67	3·52	3·37	3·21	3·05	2·87
20	8·10	5·85	4·94	4·43	4·10	3·87	3·56	3·37	3·23	3·09	2·94	2·78	2·61	2·42
30	7·56	5·39	4·51	4·02	3·70	3·47	3·17	2·98	2·84	2·70	2·55	2·39	2·21	2·01
60	7·08	4·98	4·13	3·65	3·34	3·12	2·82	2·63	2·50	2·35	2·20	2·03	1·84	1·60
∞	6·64	4·61	3·78	3·32	3·02	2·80	2·51	2·32	2·18	2·04	1·88	1·70	1·47	1·00

Answers

Note: The answers are correct, in general, to about the 3rd significant figure (small differences in the third figure are not usually of importance in statistics).

CHAPTER 2

2.3(i). 76·84, 5·12; slightly less.

2.3(ii). 67·52 in., 2·57 in.

2.3(iii). 376·11.

2.3(iv). 79·73 pence/hour; 27·34 (pence/hour)^2.

2.3(v). 257.

2.3(vi). 0·005733, 0·00089.

2.3(vii). 2,529,000.

2.4(i). 6,384, 27·34 (pence/hour)^2; 513,000, 216·7 (pence/hour)^3.

2.5(i). 38·83, 30·17; 43·56, 50·78.

CHAPTER 3

3.1(i) 0·0474, 0·0806, 0·1090, 0·2607, 0·2227, 0·1327, 0·0853, 0·0616.

3.1(ii). (a) 0·0421, 0·1311, 0·1618, 0·1553, 0·1440, 0·0987, 0·1100, 0·0680, 0·0502, 0·0210, 0·0178; (b) 0·1886, 0·0752, 0·0431, 0·0397, 0·0262, 0·0262, 0.0321, 0·0379, 0·0525, 0·0863, 0·3942; 0·6292.

3.2(ii). $p(x) = (x^2 - 2x + 5)/15$; $13/45 = 0·289$.

3.4(i). 1, 2, 0, 1.

3.4(ii). 2·5, 7, $(4^{r+1} - 1)/\{3(r + 1)\}$, 0, 0·75, $\{(1·5)^{r+1} - (-1·5)^{r+1}\}/\{3(r + 1)\}$.

CHAPTER 4

4.3(ii). (a) 0·00256; (b) 0·00069.

4.3(iii). 6 ft 0·9 in., 5 ft 4·8 in.

4.4(i). (a) 20, 112, 688, 0, 48; (b) 4·64, −11·84, 32·5888, 0, 1·2288; (c) 4,630·25, 315,707, 21,554,893, 0, 117·1875.

4.5(i). (a) 954·6; (b) 4·0.

4.5(ii). 8·2, 4·0.

4.5(iii). (a) 33·8, 2·57; (b) 14·4, 2·57; (c) −11·2, 2·57.

CHAPTER 5

5.3(i). 2·9, 1·4, 10·7, 4·06, 2·8, 29·96.

5.4(i). 3·13.
5.4(ii). 3·13.

5.5(i). 0·25, 0·0973, 0·0791 sq. in., 100, 2,500.

5.6(i). 0·02811, 0·00715.

CHAPTER 6

6.2(i). (a) Not sig.; (b) Sig. at 5%, not sig. at 1% level; (c) Sig. at 1%, not sig. at 0·1%; (d) Not sig.

6.2(ii). Not sig; not sig; sig. at 1%, not sig. at 0·1%; sig. at 1%, not sig. at 0·1%; not sig; sig. at 0·1%; not sig; sig. at 0·1%.

6.3(i). (a) Sig. at 5%, almost sig. at 2·5%; (b) Not sig.; (c) Nearly sig. at 0·1%; (d) Nearly sig. at 0·1%

6.3(ii). (a) Sig. at 5%, not sig. at 1%; (b) Sig. at 0·1% (in fact sig. at 0·05%); (c) Not sig.; (d) Sig. at 0·1% (and at 0·05%).

6.4(i). (a) Sig. at 5% not at 1%; (b) Sig. at 5%, not at 1%; (c) Sig. at 1% not at 0·1%; (d) Not sig.

6.4(ii). Sig. at 1%, not at 0·1%.

6.4(iii). Not sig.

6.4(iv). Sig. greater at 5% (and at 2·5%) but not at 1%; not sig. since $\bar{x} - \mu$ is −ve; almost sig. at 5%.

6.4(v). Not sig. as $\mu - \bar{x}$ is −ve; $\bar{x}$ is sig. lower at 1% (and 0·5%) but not at 0·1%; not sig. lower; sig. at 0·1% but not at 0·05%.

6.4(vi). Yes, at 0·1%.

6.5(i). Sig. at 1%.

6.5(ii). $c = 14·7$ so difference is very highly sig.

6.6(i). (a) Sig. at 5% level; (b) Sig. at 0·1%; (c) Nearly sig. at 1%.

6.6(ii). Sig. at 0·1%.

6.6(iii). Not sig. greater since $\bar{x} < \mu$; sig. greater at 0·05%; sig. at 1%.

6.7(i). Not sig. different.

6.7(ii). Reject null hypothesis since mean difference is sig. different from 0 at 0·1%.

6.7(iii). Accept null hypothesis since mean difference not sig. different from 0.

6.8(i). (a) Sig. at 5%; (b) Nearly sig. at 1%; (c) Not sig; (d) Not sig.

6.8(ii). Sig. at 0·1%.

6.8(iii). Sig. greater at 1%; sig. greater at 5%; not sig. greater.

6.8(iv). (a) Not sig. different since obs $W = 1·87 <$ crit $W = 2·08$ at 5%; no sig. difference since obs $W = 0·54$; obs $W = 4·23$, crit $W = 4·14$ at 0·1%, the degrees of freedom for $t = 13·9$; (b) Not sig. greater since $\bar{x} < \bar{x}'$; not sig. greater; sig. greater.

6.9(i). (*a*) Sig. at 5% not at 1%; (*b*) Sig. at 5 not at 1%; (*c*) Not sig.

6.9(ii). Not sig.; not sig.; not sig.

CHAPTER 7

7.2(i). $\bar{x}_1, \ldots, \bar{x}_5 = 8,\ 8,\ 5,\ 14,\ 5$; $\bar{x} = 8$; $(\bar{x}_1 - \bar{x})^2 + \ldots + (\bar{x}_5 - \bar{x})^2 = 0 + 0 + 9 + 36 + 9 = 54$; $\{(x_{11} - \bar{x}_1)^2 + (x_{12} - \bar{x}_1)^2 + (x_{13} - \bar{x}_1)^2\} + \{(x_{21} - \bar{x}_2)^2 + (x_{22} - \bar{x}_2)^2 + \ldots + (x_{53} - \bar{x}_5)^2\} = \{0 + 1 + 1\} + \{4 + 0 + 4\} + \{4 + 1 + 1\} + \{1 + 4 + 1\} + \{16 + 1 + 9\} = 48$; $(s')^2 = 54/(5 - 1) = 13.5$; $(s'')^2 = 48/\{5(3 - 1)\} = 4.8$; hence $n(s')^2/(s'')^2 = 3 \times 13.5/4.8 = 8.44 > \text{crit } F(\nu_1 = 4, \nu_2 = 10) = 5.99$ at 1%, so difference between means is sig. at 1%.

7.3(i). $n(s')^2/(s'')^2 = 12.85/3.24 = 3.98 > \text{crit } F(\nu_1 = 2, \nu_2 = 15) = 3.68$ at 5%; $\sigma^2 = 3.24$, $\sigma_b^2 = 1.069$.

7.3(ii) $n(s')^2/(s'')^2 = 3.05/0.756 = 4.04 > F(\nu_1 = 8, \nu_2 = 36) = 3.05$ at 1%; $\sigma^2 = 0.756$, $\sigma_b^2 = 0.459$.

7.3(iii). $n(s')^2/(s'')^2 = 52/5 = 10.4 > F(\nu_1 = 3, \nu_2 = 8) = 7.59$ at 1%; $\sigma^2 = 5.00$, $\sigma_b^2 = 15.67$.

7.3(iv). $n(s')^2/(s'')^2 = 34.5/5.6 = 6.16 > F(\nu_1 = 4, \nu_2 = 10) = 5.99$ at 1%; $\sigma^2 = 5.6$, $\sigma_b^2 = 9.63$.

7.4(i). $V_b/V_r = 214.7/20.67 = 10.4 > F(\nu_1 = 5, \nu_2 = 114) = 3.18$ at 1%; $\sigma^2 = 20.67$, $\sigma_b^2 = 10.35$.

7.5(i). σ_1^2 sig. at 5% not at 1%; σ_2^2 sig. at 1%; $\sigma^2 = 0.375$, $\sigma_2^2 = 1.544$.

7.5(ii). σ_1^2 sig. at 1%, σ_2^2 sig. at 1%; $\sigma_1^2 = 0.304$, $\sigma_2^2 = 0.716$.

7.5(iii). σ_1^2, σ_2^2 sig. at 1%; $\sigma_1^2 = 16.89$, $\sigma_2^2 = 3.67$.

7.5(iv). σ_1^2, σ_2^2 sig. at 1%; $\sigma_1^2 = 10.83$, $\sigma_2^2 = 3.60$.

CHAPTER 8

8.3(i). (*a*) 6.13, 8.57; 5.74, 8.96; 5.30, 9.40; (*b*) −21.11, +11.45; −26.26, +16.60; −32.16, +22.50; (*c*) 78.35, 84.45; 77.38, 85.42; 76.28, 86.52.

8.4(i). (*a*) 9.0, 25.8; 5.3, 29.5; −0.4, 35.2; (*b*) −33.5, −9.7; −39.5, −3.7; −50.4, +7.2;

(*c*) −6.95, −3.19; −7.63, −2.51; −8.50, −1.64.

8.4(ii). 5.74, 13.29; 3.59, 15.43; −0.57, 19.59.

8.4(iii). −5.94, 11.58; −11.67, 17.31; −24.30, 29.94.

8.4(iv). Reject 32.00 and 27.28 at the 0.1% level.

8.5(i). For the data of 6.8(i) (a) the 95, 99, 99·9% limits are −17·8, −0·8; −20·9, +2·3; −24·8, +6·2; of 6·8(i) (b), −0·104, −0·012; −0·119, +0·001; −0·136, +0·018; of 6.8(i) (c), −17·0, +1·0; −20·4, +4·4; −24·9, 8·9; of 6.8(i) (d), −5·6, 9·0; −8·1, 11·5; −11·4, 14·8; for data of 6.8(ii), 3·09, 9·63; 1·83, 10·89; 0·12, 12·60.

8.5(ii). −16·9, 0·9; −4·7, 8·1.

8.6(i). (a) 4·18, 64·4; 3·20, 129·9; (b) 46·1, 290·8; 37·5, 437·5; (c) 0·0359, 0·158; 0·0301, 0·215; (d) 5·06, 77·9; 3·88, 157; (e) 17·9, 410; 13·3, 959; (f) 0·0133, 0·0937; 0·0110, 0·1463; (2nd sample) 0·00349, 0·0220; 0·00284, 0·0331.

8.6(ii). (a) Accept at 5%; (b) Accept at 5%; (c) Accept at 5%; (d) Reject at 1%, since 1% limits are 72·2, 157·4.

CHAPTER 9

9.2(i). (a) $\frac{1}{12}$; (b) $\frac{1}{6}$.

9.2(ii). (a) $\frac{1}{864}$; (b) $\frac{1}{108}$.

9.2(iii). (a) $\frac{22}{52}$; (b) $\frac{18}{52}$.

9.2(iv). (a) $\frac{3}{13}$; (b) $\frac{1}{2}$.

9.2(v). $\frac{5}{13}$.

9.3(i). (a) 250/7,776 = 0·03215; (b) 4,375/279,936 = 0·01563.

9.3(ii). (a) $\frac{1}{9}$; (b) $\frac{1}{12}$; (c) $\frac{1}{6}$.

9.4(i). (a) $\frac{15}{64}$; (b) $\frac{21}{128}$.

9.4(ii). (a) $36/13^3 = 36/2197$; (b) $3^3 10^3/13^5 = 27,000/371,293$.

9.4(iii). 0·32768, 0·40960, 0·00032.

9.5(iii). 0·0605, 0·0880, 0·0326 (approx).

9.5(iv). (a) 4·2, 1·26; (b) 1·2, 1·020; (c) 2, 1·96.

9.5(v). 0·02849, 0·02515, 0·01729, 0·02515, 0·01729, 0·00925.

9.7(i). (a) Accept the hypothesis that p could be $\frac{1}{6}$; (b) reject at 5%, but accept $p = \frac{1}{6}$ at 1%.

9.7(ii). p could be $\frac{1}{6}$, but results are nearly sig. at 5%.

9.7(iii). Probability of 0, 1 successes are $^6C_0(\frac{9}{10})^6$, $^6C_1(\frac{1}{10})(\frac{9}{10})^5 = 0·531$, 0·354, hence probability of 2 or more is $1 − 0·531 − 0·354 = 0·115$, so result is not sig. at 5% level.

9.8(i). (a) 0·197, 0·150; 0·205, 0·143; (b) 0·282, 0·150; 0·308, 0·135; (c) 0·138, 0·058; 0·156, 0·050.

9.8(ii). (a) 11,111; 14,800, 8,380; 16,200, 7,680; (b) 2,143; 3,075, 1,505; (c) 153,060; 175,800, 133,300; 183,700, 127,600.

9.8(iii). (a) 0·0124, 0·144; 0·0068, 0·185; (b) 0·0136, 0·1094; 0·0084, 0·1375; (c) 0·0163, 0·1025; 0·0108, 0·1260; (d) 0·0240, 0·0855; 0·0186, 0·1000; (e) 0·0036, 0·0292; 0·0022, 0·0367.

9.8(iv). 0·126, 0·168; 0·097, 0·126; 0·0915, 0·115; 0·0785, 0·0940; 0·0258, 0·0335.

9.8(v). Given hypothesis $p = 0.13$ then for data of 9.8(i) (a) reject at 1%; of 9.8(i)(b) reject at 1%; of 9.8(i) (c) accept at 5%; given hypothesis $p = 0.25$ then for data of 9.8(i) (a) reject at 1%; of 9.8(i) (b) accept at 5%; of 9.8(i) (c) reject at 1%.

9.9(i) (95% confidence intervals given first). (a) 0.162, 0.044, $\lambda' = 43.1$, $\lambda'' = 87.45$; 0.194, 0.036, $\lambda' = 50.9$, $\lambda'' = 127.8$; (b) 0.173, 0.047, $\lambda' = 38.9$, $\lambda'' = 85.4$; 0.211, 0.039, $\lambda' = 44.6$, $\lambda'' = 124.8$; (c) 0.290, 0.144, $\lambda' = 33.4$, $\lambda'' = 52.1$; 0.320, 0.129, $\lambda' = 40.85$, $\lambda'' = 73.3$.

9.9(ii). (a) 0.153, 0.041, $\lambda' = 45.85$, $\lambda'' = 93.6$; 0.184, 0.034, $\lambda' = 54.0$, $\lambda'' = 136.7$; (b) 0.111, 0.043, $\lambda' = 82.2$, $\lambda'' = 138.7$; 0.129, 0.037, $\lambda' = 99.5$, $\lambda'' = 197.4$.

9.9(iii). For data of 9.9(i) (a) 0.147, $\lambda' = 38.3$, 0.181, $\lambda' = 48.0$; of 9.9(i) (b) 0.156, $\lambda' = 35.1$, 0.195, $\lambda' = 42.6$; of 9.9(i) (c) 0.275, $\lambda' = 29.1$, 0.308, $\lambda' = 38.0$; of 9.9(ii) (a) 0.138, $\lambda' = 40.7$, 0.171, $\lambda' = 51.0$; of 9.9(ii) (b) 0.103, $\lambda' = 72.0$, 0.121, $\lambda' = 92.9$.

9.10(i). $175/11{,}664 = 0.01500$.

CHAPTER 10

10.2(i). (a) If probability $= 0.02$ then probability of 0, 1, 2, 3, 4 *in* 1 *min* is 0.3012, 0.3614, 0.2169, 0.0867, 0.0260; *in* 100 *sec* is 0.1353, 0.2707, 0.2707, 0.1805, 0.0902; *in* 5 *min* is 0.0025, 0.0149, 0.0446, 0.0893, 0.1339; (b) If probability $= 0.005$, then probability of 0, 1, 2, 3, 4 *in* 1 *min* is 0.7408, 0.2222, 0.0333, 0.0033, 0.0003; *in* 100 *sec* is 0.6065, 0.3033, 0.0758, 0.0126, 0.0016; *in* 5 *min* is 0.2231, 0.3347, 0.2510, 0.1255, 0.0471.

10.2(ii). (a) With $\mu = 1.3225$, $e^{-\mu} = 0.26647$, so theoretical frequencies are 106.6, 141.0, 93.2, 41.1, 13.6, 3.6, 0.8, 0.1; (b) $\mu = 1.8000$, so $e^{-\mu} = 0.16530$, and theoretical frequencies are 66.1,

119.0, 107.1, 64.3, 28.9, 10.4, 3.1, 0.8, 0.2; (c) $\mu = 4.6958$, $e^{-\mu} = 0.0091334$, so theoretical frequencies are 3.7, 17.2, 40.4, 63.2, 74.2, 69.7, 54.5, 36.6, 21.5, 11.2, 5.3, 2.2, 0.9, 0.3.

10.3(i). (a) $\Pr(s) = 0.04755$, 0.14707, 0.22515, 0.22799, 0.17099; (b) $\Pr(s) = 0.04979$, 0.14936, 0.22404, 0.22404, 0.16803.

10.3(ii). (a) 0.4404, 0.3612, 0.1481, their sum $= 0.9497 = 0.95$, and $\chi^2_{0.95}$ $(\nu = 6) = 1.64 = 2 \times 0.82$. (b) 0.3362, 0.3665, 0.1997, 0.0726, sum $= 0.9750$, $\chi^2_{0.975}(\nu = 8) = 2.18 = 2 \times 1.09$; (c) 0.3396, 0.3668, 0.1981, 0.0713, 0.0193, sum $= 0.9951$, $\chi^2_{0.995}(\nu = 10) = 2 \times 1.08$.

CHAPTER 11

11.2(i). $\chi^2 = 8.32$, $\chi^2_{0.05}(\nu = 5) = 11.1$, so accept $p = \frac{1}{6}$ hypothesis.

11.2(ii). $\chi^2 = 0.052 < \chi^2_{0.99}(\nu = 3) = 0.115$.

11.4(i). Uncorrected $\chi^2 = 4$, corrected $\chi^2 = 3.61$, $\chi^2_{0.05}(\nu = 1) = 3.84$, so borderline case.

11.4(ii). Corrected $\chi^2 = 6.25$, nearly sig. at 1% so reject $p = \frac{1}{2}$ hypothesis.

11.5(i). Yes, corrected $\chi^2 = 28.27 \gg \chi^2_{0.001}$ $(\nu = 1) = 10.8$.

11.5(ii). $\chi^2 = 5.93$, so reject no association hypothesis at 5% but not at 1%.

11.5(iii). Corrected $\chi^2 = 24.45$, $\chi^2_{0.01}(\nu = 8) = 20.1$ so evidence of difference is very strong.

11.6(i). No individual χ^2 sig., but Total $\chi^2 = 10.85 > \chi^2_{0.05}$ $(\nu = 4) = 9.45$ so sig. at 5%.

11.7(i). (a) For Table 10.1 data group $s = 2, 3, 4, \ldots$;

uncorrected $\chi^2 = 0.51 < 3.84 = \chi^2_{0.05}(\nu = 1)$ so fit is good; (b) for data of 10.2(ii)(a) group $s = 4, 5, 6 \ldots$; uncorrected $\chi^2 = 1.36 < \chi^2_{0.05}$ $(\nu = 2) = 5.99$ so fit is good; for data of 10.2(i) (b) group $s = 5, 6, \ldots$; uncorrected $\chi^2 = 7.24 < 9.49 = \chi^2_{0.05}(\nu = 4) = 9.49$ so good fit; for data of 10.2(i)(c) group $s = 0$, 1; and group $s = 9, 10, \ldots$; $\chi^2 = 4.30 < 14.1 = \chi^2_{0.05}(\nu = 7)$ so good fit.

11.7(ii). (a) Group $s = 4$, 5, $\ldots$; $\chi^2 = 1.237 < 9.49 = \chi^2_{0.05}$ so no reason to reject hypothesis.

11.7(iii). Put $\mu = 1.434 =$ mean observed frequency, hence theoretical frequencies are 34.11, 48.90, 35.05, 16.75, 6.00, 1.72, 0.41; group $s = 4, 5, \ldots$ then $\chi^2 = 1.32 < \chi^2_{0.05}(\nu = 4) = 9.49$ so very good fit.

11.7(iv). $\chi^2 = 2.90 < \chi^2_{0.05}$ $(\nu = 3) = 7.81$ so good fit.

CHAPTER 12

12.2(i). (a) 0.9, -0.8, 2.8, 0.1, $S = 9.30$; (b) 0.7, -1.4, 1.4, -1.7, $S = 7.30$; (c) 0.55, -1.35, -1.85, -1.05, $S = 6.65$.

12.2(ii). (a) $y = 1.349x + 0.516$; (b) $y = -0.88x + 10.88$.

12.3(i). (a) $y = (97/20) + (7/120)x + (11/24)x^2 = 4.85 + 0.0583x + 0.4583x^2$; $y = (4,287/870) + (113/348)x + (73/174)x^2 - (3/116)x^3 = 4.928 +$

$0.325x + 0.420x^2 - 0.026x^3$; (b) $y = -9.461 + 6.550x - 0.669x^2$; $y = -7.021 + 3.750x + 0.0831x^2 - 0.0557x^3$.

12.4(i). (a) $y = 3.973x - 0.4626x^2$; (b) $y = 0.5695x + 7.906/x$.

12.5(i). Least square line is $y = 1.00 - 0.25x$, coefficient of determination $= 0.6875/34 =$

0·0202; Least square quadratic curve is $y = -6·286 + 4·607x -0·6071x^2$, coefficient of determination $= 32·214/34 = 0·9475$.

12.6(i). For data of 12.2(ii) (a) obs $F = 70·3 >$ crit $F(\nu_1 = 1, \gamma_2 = 4) = 21·2$ at 1%, so highly sig.; for data of 12.2(ii)(b) obs $F = 47·8 > 21·2$ so highly sig.; for data of 12.3(i)(a) obs $F = 0·47 < 1$ so not sig.; for data of 12.3(i)(b) obs $F = 0·073$ so not sig.

12.6(ii). For data of 12.3(i) (a) $S_1^2/S_2^2 = 17·525/0·075 = 233·7 \gg F(\nu_1 = 2, \nu_2 = 2) = 9·9$ at 1%; for data of 12.3(i)(b) $S_1^2/S_2^2 = 41·1/2·154 = 19·9 > F$ $(\nu_1 = 2, \nu_2 = 5) = 12·06$ at 1%, so quadratic curve is better than no relation in both cases.

12.7(i). (a) $y = 1·40x + 0·20$; (b) $y = -x + 11$.

12.8(i). $z = 10·87 + 2·641x -3·685y$.

CHAPTER 13

13.2(i). (a) $-7·25$; (b) $+7·7$; (c) $-2·583$.

13.2(ii). (a) $+ 0·500$; (b) $-0·8705$; (c) $-0·115$.

13.2(iii). (a) $-0·899$; (b) $+0·988$; (c) $-0·189$.

13.4(i). For data of 13.2(i)— y on x line is
(a) $y - 8 = -1·115(x - 4)$
(b) $y - 12 = 0·643(x - 7·5)$
(c) $y - 6·75 = -0·29(x + 0·75)$
x on y line is—
(a) $y - 8 = -1·375(x - 4)$
(b) $y - 12 = 0·658(x - 7·5)$
(c) $y - 6·75 = -8·10(x = 0·75)$

13.5(i). (a) $c = 1·29$, not sig., so no evidence of correlation; (b) $c = 2·59$ sig. at 1% so reject $\rho = 0$, i.e. there is some correlation; (c) $c = 0·357$ not sig. so no evidence of correlation.

13.5(ii). (a) $c = 3·35$, sig. at 1% so reject $\rho = 0·30$; (b) $c = 0·93$, not sig., so ρ could be $-0·40$.

13.7(i). $\rho_s = 31/42 = 0·738$; $t_s = 2·68 > t(\nu = 6) = 2·45$ at $2·5\%$ (one-tailed) so agreement (concordance) is good.

13.7(ii). For data of 13.2(i) (a) $\rho_s = 0·9, t_s = 3·576 > t(\nu = 3) = 3·18$ at 5% (two-tailed); for data of 13.2(i) (b) give the equal y-values rank $2·5$, $\rho_s = 0·986$, $t_s = 11·70 > t(\nu = 4) = 8·61$ at $0·1\%$ so association is very strong; (c) $\rho_s = 0 = t_s$, so association is very weak.

CHAPTER 14

14.2(i). (a) Difference between samples sig. at $0·5\%$; (b) Difference sig. at 5% but not at $2·5\%$.

14.3(i). For data of 14.2(i), $M = 8·17 > \chi^2_{0·005}(\nu = 1) =$ 7·88 so difference between medians is sig. at $0·5\%$; of 14.2(i) (b); $M = 0$ so difference between medians is not sig. (the dispersions are different see 14.2(i) (b) above).

14.3(ii). (a) $M = 0.67$ not sig.; (b) $M = 2.69$ not sig.; (c) $M = 6.04$ sig. at 2.5%.

14.3(iii) (a) $M = 8.16$, sig. at 0.5%; (b) $M = 4.17$, sig. at 5% not at 2.5%; (c) Subtract 22.9, then $M = 4.17$, sig. at 5%, subtract 23.1 and again $M = 4.17$.

14.3(iv). 2 ≤ (2nd median— 1st median) ≤ 25.

14.4(i). $S = 0.1$, not sig.

14.4(ii). (a) $S = 1.56$ not sig.; (b) $S = 3.29$ not sig.

14.4(iii). 95% limits are +4, −4; 99% limits are +9, −6 (k being subtracted from the first observation of each pair).

14.4(iv). 95% limits are 8, 26; 99% limits are 0, 30.

14.5(i). (a) No sig. difference; (b) Sig. at 5% not at 1%.

CHAPTER 16

Here, cr means critical, p being the corresponding percentage probability level (for $\chi^2_{0.995}$, $\chi^2_{0.001}$ then $p = 99.5$, or 0.1); g stands for grouped; in Examples 16.1–3, $W = |\mu - \bar{x}'|/\sqrt{\{s^2/n + (s')^2/n'\}}$ and the answer to (g) is 'yes', since $W >$ cr W.

	(b) h	μ	σ²	σ	(c) σ²	(d) x̄	s	(f) x̄′	s′	(g) W	cr W	p
16.1	0·050	32·050	0·015625	0·125	0·01542	32·050	0·127	32·150	0·0736	3·4	2·82	1
16.2	0·50	45·500	2·2525	1·50	2·2504	45·484	1·48	46·40	0·595	3·2	2·72	1
16.3	1·00	34·500	2·000	1·41	1·9167	34·470	1·54	33·00	1·095	3·7	2·92	1

16.4. No: $W = 1.4 <$ cr $W = 2.07$, $p = 5$.

16.5. Yes: $W = 3.2 >$ cr $W = 2.60$, $p = 1$.

16.6. Yes: $W = 3.0 >$ cr $W = 2.28$, $p = 5$.

In what follows good* means 'too good', i.e. the readings may be 'cooked'; c means corrected and u means uncorrected.

	μ	Expected frequencies	χ²	crχ²	ν	p	Fit
16.7	0·46522	406, 189, g7·7	c.62·6	13·8	4 − 2 = 2	0·1	bad
16.8(A)	3·9	2·0, 7·9, 15·4, 20·0, 19·5, 9·9, 5·5, g4·54	u0·22	0·99	9 − 2 = 7	99·5	good*
(B)	3·0	5·0, 14·9, 22·4, 16·8, 10·1, g8·4	u0·21	0·41	7 − 2 = 5	99·5	good*
(i) (C)	1·8	16·5, 29·8, 16·1, g10·9	c11·3	11·3	5 − 2 = 3	1	bad
(D)	2·4	18·1, 1, 43·6, 52·3, 41·8, 12·0, g7·1	c16·5	15·1	7 − 2 = 5	1	bad
(ii) D	1·6, 3·2	24·3, 45·4, 46·7, 36·1, 23·3, 13·2, g11·1	u3·85	7·8	7 − 3 = 4	5	good

16.8(iii). A and B, and C and D both significant difference, $p = 1$, but D is not true Poisson.

16.9. No, $u\chi^2 = 1.38 <$ cr χ^2 $(\nu = 8 - 5) = 7.8, p = 5$.

16.10. (a) Yes, $u\chi^2 = 182 >$ cr χ^2 $(\nu = 9 - 5) = 18.5, p = 0.1$; (b) Yes, $u\chi^2 = 171 >$ cr $\chi^2(\nu = 6 - 4) = 13.8, p = 0.1$; (c) yes, $u\chi^2 = 68 > 13.8, p = 0.1$; (d) No, $u\chi^2 = 4.5 <$ cr χ^2 $(\nu = 6 - 4) = 5.99, p = 5$.

16.11. (a) Yes, $c\chi^2 = 5.0 >$ cr $\chi^2(\nu = 1) = 3.84, p = 5$; (b) some types of bowling may produce more l.b.w. decisions than others.

16.12. In each case fit is good and expected frequencies are: g5.75, 4.84, 6.64, 7.77, 7.77, 6.64, 4.84, g5.75; $u\chi^2 = (a)$, 2.9, (b) 5.3, (c) 3.7 < cr $\chi^2(\nu = 8 - 3) = 11.1, p = 5$.

16.13.

	μ	σ	Expected frequencies	χ^2	crχ^2	ν	p	Fit
(i)	981	2.00	g4.6, 8.8, 18, 30, 38, 38, 30, 18, 8.8, g4.6	u8.2	14		7 5	good
(ii)	40	100	g8.1, 7.8, 11.6, 14.6, 15.7, 14.6, 11.6, 7.8, g8.1	u2.4	14		7 5	good
(iii)	20.4	5.0	g4.6, 8, 17, 32, 50, 68, 77, 65, 46, 29, 15, g10	c29	29.6		10 0.1	bad
(iv)	184.5	18.8	g6.5, 9.6, 18, 29, 38, 42, 39, 30, 19, 10.5, g7.3	u3.1	15.5		8 5	good
(v)	11.85	3.42	g22, 18, 27, 37, 48, 56, 60, 59, 53, 45, 34, 24, 16, 9, g9	c58	32.9		12 0.1	bad
(vi)	1.50	0.50	–, 6, 19, 56, 132, 244, 352, 398, 352, 244, 132, 56, 19, 6,–	c76	29.6		10 0.1	bad

$(A) |\mu - 0|/(\sigma/\sqrt{n}) = 40/(100/10) = 4 > 3.29$ so significant difference, $p = 0.1$.

16.14. Good fit in each case; (a) $u\chi^2 = 12 <$ cr $\chi^2(\nu = 15 - 5) = 18.3, p = 5$; (b) $u\chi^2 = 11 <$ cr $\chi^2(\nu = 11 - 5) = 12.6, p = 5$; further details are:

μ_1, σ_1, H_1; μ_2, σ_2, H_2; Expected frequencies =
(a) 10.3, 2.3, 72; 16.2, 1.6, 41; g11, 17, 31, 47, 60, 63, 56, 43, 34, 33, 38, 37, 27, 13, g5.9; (b) 1.10, 0.305, 305; 1.85, 0.325, 335; –, 6.3, 44, 158, 300, 331, 315, 351, 305, 157, 45, 7.2. –.

16.15. (a) for (A), $r = 0.854$, $y = 0.569x + 1.208$, $x = 1.282y + 4.735$; for (B), $r = 0.830$, $y = 0.192x + 2.523$, $x = 3.584y - 14.435$; for (C), $r = 0.919$, $y = 0.588x - 2.045$, $x = 1.437y + 4.685$. (b) for (A), $r\sqrt{(N - 1)} = 4.0 > 3.29$, so r significantly different from 0, $p = 0.1$, see (13.13); $t_2(\beta_2' = 0) = 7.5 >$ cr $t(\nu = 21) = 3.8, p = 0.1$, so β_2, β_1 significantly different from 0, see (16.1); for (B), $r\sqrt{(N - 1)} = 2.35 > 1.96$, so r significantly different from 0, $p = 5$; $t_2(\beta_2' = 0) = 3.93 >$ cr $t(\nu = 7) = 3.36$, $p = 1$, so β_2, β_1 significantly different from 0; for (C), $r\sqrt{(N-1)} = 3\cdot8 > 3\cdot29$ so r significantly different from 0, $p = 0\cdot1$; $t_2(\beta_2' = 0) = 9\cdot3 > $ cr $t(\nu = 16) = 4\cdot14$, $p = 0\cdot1$ so β_2, β_1 significantly different from 0. (c) $U = 0\cdot03$ so r_A, r_B not significantly different; the β_2's are significantly different, as $(S')^2/S^2 = 3\cdot3/0\cdot232 = 14\cdot2 > F(\nu_1 = 1, \nu_2 = 30) = 7\cdot56$, $p = 1$; the β_1's are significantly different as $(S')^2/S^2 = 10\cdot9/0\cdot907 = 12\cdot0 > F(\nu_1 = 1, \nu_2 = 30) = 7\cdot56$, $p = 1$. (d) $\bar{r} = 0\cdot848$, $\bar{\beta}_2 = 0\cdot400$, $\bar{\beta}_1 = 1\cdot487$. (e) Reject common y on x line as $Z^2/S^2 = 5\cdot12/0\cdot232 = 22 > F(\nu_1 = 2, \nu_2 = 30) = 5\cdot4$, $p = 1$; reject common x on y line as $Z^2/S^2 = 38\cdot6/0\cdot907 = 43 > 5\cdot4$.

16.16. (A) $r = 0\cdot992$, $y = 0\cdot675x - 38\cdot18$, $x = 1\cdot458y + 61\cdot12$; (B) $r = 0\cdot847$, $y = 0\cdot609x + 115\cdot5$, $x = 1\cdot177y + 66\cdot2$; (C) $r = 0\cdot843$, $y = 0\cdot478x + 315\cdot9$, $x = 1\cdot490y - 216\cdot8$. (a) $r\sqrt{(N-1)}$, for (A) $= 2\cdot43$, for (B) $= 2\cdot45$, for (C) $= 2\cdot07$, so all r significantly different from 0, $p = 5$; $t_2(\beta_2' = 0)$ for (A) $= 17\cdot6 > $ cr $t(\nu = 5) = 6\cdot9$, $p = 0\cdot1$, for (B) $= 3\cdot6 > $ cr $t(\nu = 5) = 2\cdot6$, $p = 5$, for (C) $= 3\cdot5 > 2\cdot6$, $p = 5$, so all β_2, β_1 significantly different from 0. (b) $\bar{r} = 0\cdot941$, $\bar{\beta}_2 = 0\cdot587$, $\bar{\beta}_1 = 1\cdot345$. (c) $U = 6\cdot17 > $ cr $\chi^2(\nu = 2) = 5\cdot99$, $p = 5$ so r's significantly different; no significant difference between β_2's as $(S')^2/S^2 = 1720/3031 < 1$; no significant difference between β_1's as $(S')^2/S^2 = $ 2480/7142 < 1. (d) Reject common y on x line as $Z^2/S^2 = 5\cdot5 > F(\nu_1 = 4, \nu_2 = 15) = 4\cdot9$, $p = 1$. (e) Accept common x on y line as $Z^2/S^2 = 7425/7142 = 1\cdot04 < F(\nu_1 = 4, \nu_2 = 15) = 3\cdot06$, $p = 5$. (f) The y on x levels significantly different as $(Z')^2/(S'')^2 = 31,430/2,877 = 10\cdot9 > F(\nu_1 = 2, \nu_2 = 17) = 6\cdot1$, $p = 1$. (g) Accept common x on y level as $(Z')^2/(S'')^2 = 12,370/6,594 = 1\cdot88 < F(\nu_1 = 2, \nu_2 = 17)$, $p = 5$.

16.17. $r = 0\cdot694$, $y = 0\cdot412x + 82\cdot2$, $x = 1\cdot168y - 63\cdot4$. $r\sqrt{(N-1)} = 2\cdot94 > 2\cdot58$, so r significantly different from 0, $p = 1$; $t_2(\beta_2' = 0) = 3\cdot97 > $ cr $t(\nu = 17) = 3\cdot96$, $p = 1$ so β_2, β_1 significantly different from 0. Quadratic curve is $y = 83\cdot0 + 0\cdot3873x + 0\cdot0001857x^2$, $S_1^2 = 323\cdot3$, (s_1^2 of (12.16) $= C^2/X$, here), so $2S_1^2 - s_1^2 < 0$, so quadratic curve no improvement over line.

16.18. $r = 0\cdot759$, $y = 0\cdot0824x + 40\cdot68$. $r\sqrt{(N-1)} = 3\cdot7 > 3\cdot29$, so r significantly different from 0, $p = 0\cdot1$; $t_2(\beta_2' = 0) = 5\cdot6 < $ cr $t(\nu = 23) = 3\cdot77$, $p = 0\cdot1$, so β_2 significantly different from 0, hence there is change with temperature. Quadratic curve $y = 43\cdot08 + 0\cdot1104(x - 30) - 0\cdot001495(x - 30)^2$ shows no improvement over line.

16.19. $\lambda = -0\cdot2280$, 3·04 hours, 645·3.

16.20. $\lambda = -0\cdot01282$, 54·07 hours, 3609.

Example	Data of	ρ_s	t_s	cr t	ν	p	Significantly different from 0?	$\rho = 2\sin(30\rho_s{}^0)$
16.22	(i)	0·397	1·22	2·31	8	5	No	—
	(ii)	0·758	4·33	3·36	8	1	Yes	—
	(iii)	0·579	2·01	2·31	8	5	No	—
16.23	16.15(A)	0·806	6·24	3·82	21	0·1	Yes	0·819
	16.15(B)	0·771	3·11	2·36	7	5	Yes	0·785
	16.15(C)	0·879	7·38	4·01	16	0·1	Yes	0·889
	16.16(A)	0·929	5·59	4·03	5	1	Yes	0·934
	16.16(B)	0·857	3·72	2·57	5	5	Yes	0·868
	16.16(C)	0·893	4·43	4·03	5	1	Yes	0·901
	16.17	0·635	3·39	2·90	17	1	Yes	0·653
	16.18	0·745	5·36	3·77	23	0·1	Yes	0·761

16.21. (a) $k = 0·3017$; (b) 23·0 min; (c) 23·0 min.

16.24(i). var $g = \{2T/(4\pi^2L)\}^2$ var $T + \{T^2/(4\pi^2L^2)\}^2$ var L.

16.24(ii). var $T = 0·0822$, var $L = 0·000244$; $S = 0·01413$. 14,000 : 1.

16.25. m, $S = 0·0018, 0·000919$ (a); 0·085, 0·00216 (b); 0·0832, 0·00235 (c); 0·0416, 0·00117 (d); 0·005437, 0·000307 cm² (e) 0·3002, 0·01695 cm³ (f); 9·277, 0·525 g/cm³ (g), a large standard error (wire is not the best shape for density measurements).

16.26. $m, S = 2·0002, 0·00148$ cm (a); 4·1913, 0·00927 cm³ (b); 11·293, 0·02499 g/cm³ (c); (a sphere is better than wire).

16.27. (a) var $\eta = (\cos i/\sin r)^2$ var $i + (\sin i \cos r/\sin^2 r)^2$ var r; (b) 1·5208, 0·0161; (N.B. var i, var r, etc. must be in radians).

16.28. (a) Yes, see (7·11), $V_2/V_r = 0·96/0·1067 = 9·0 > F(\nu_1 = 2, \nu_2 = 6) = 5·14, p = 5$; (b) Yes, $V_1/V_r = 0·72/0·1067 = 6·75 > F(\nu_1 = 3, \nu_2 = 6) = 4·8$, $p = 5$.

16.29. (a) Yes, $V_1/V_r = 82/29 = 2·83 > F(\nu_1 = 5, \nu_2 = 20) = 2·7, p = 5$; (b) Yes, $V_2/V_r = 84/29 = 2·90 > F(\nu_1 = 4, \nu_2 = 20) = 2·87, p = 5$.

16.30. (a) Yes, $V_1/V_r = 34,360/2,308 = 14·9$. $F(\nu_1 = 4, \nu_2 = 16) = 5·29, p = 1$; (b) Yes, $V_2/V_r = 23,512/2,308 = 10·2 > 5·29$.

APPENDIX

A(i). RO, RY, RG, RB, OY, OG, OB, YG, YB, GB; no. of combinations = 10; RGB, ROY, ROG, ROB, RYG, RYB, OYG, OYB, OGB, YGB; No. of combinations = 10; ROYG, ROYB, ROGB, RYGB, OYGB; No. of combinations = 5.

A(ii). 10, 10, 5.

A(iii). 35, 45, 9, 70.

A(iv). 120; 5,040; 40,320.

A(v). 1, 15, 15, 6, 45, 120, 1, 12.

A(vi). $p^3 + 3p^2q + 3pq^2 + q^3$; $p^5 + 5p^4q + 10p^3q^2 + 10p^2q^3 + 5pq^4 + q^5$; $a^4 + 4a^3b + 6a^2b^2 + 4ab^3 + b^4$.

Index

171